The Multifaceted Nature of Creativity in the Teaching of Geometry

The Multifaceted Nature of Creativity in the Teaching of Geometry

DORIT PATKIN
Kibbutzim College of Education Technology & Arts,
Tel-Aviv, Israel

ATARA SHRIKI
Kibbutzim College of Education Technology & Arts,
Tel-Aviv, Israel

ILANA LEVENBERG
Gordon-Academic College of Education,
Haifa, Israel

World Scientific

NEW JERSEY • LONDON • SINGAPORE • BEIJING • SHANGHAI • HONG KONG • TAIPEI • CHENNAI • TOKYO

Published by

World Scientific Publishing Co. Pte. Ltd.
5 Toh Tuck Link, Singapore 596224
USA office: 27 Warren Street, Suite 401-402, Hackensack, NJ 07601
UK office: 57 Shelton Street, Covent Garden, London WC2H 9HE

Library of Congress Control Number: 2020042483

British Library Cataloguing-in-Publication Data
A catalogue record for this book is available from the British Library.

Spine cover photo by S. Kirsner

THE MULTIFACETED NATURE OF CREATIVITY IN THE TEACHING OF GEOMETRY

ISBN 978-981-121-874-3 (hardcover)
ISBN 978-981-121-875-0 (ebook for institutions)
ISBN 978-981-121-876-7 (ebook for individuals)

For any available supplementary material, please visit
https://www.worldscientific.com/worldscibooks/10.1142/11781#t=suppl

https://doi.org/10.1142/9789811218750_fmatter

Acknowledgment

We wish to thank all of our colleagues who took part in writing some of the chapters in this book.

Special thanks go to the professional and devoted linguistic editor of the book, Ms. Nira Ben Ari, who invested a lot of thought and work in editing and designing the entire book.

We also wish to thank Sarah Gazit and Shai Kirsner for the spectacular photos included in this book and to our dedicated desk editor Tan Rok Ting.

Finally, on a personal note from us, the Editors: This book is dedicated with great love to all our grandchildren, the next generation of creators. We hope that creative mathematical thinking and the joy of creation, will always accompany you and enlighten your way.

Dorit Patkin, Atara Shriki, and Ilana Levenberg

https://doi.org/10.1142/9789811218750_fmatter

Table of Contents

https://doi.org/10.1142/9789811218750_fmatter

Prolog

You can't use up creativity. The more you use, the more you have!

Oscar Wilde

This book discusses creativity in the teaching of geometry. Before delving into the content of the book, we wish to relate briefly to its key aspects: creativity and geometry, while establishing a relation between them.

On creativity, its Role and Nurturing at School

If you were asked to name some well-known people who you thought were creative, what names would come to mind? What were your criteria for choosing those people? Were these people from the area of painting? sculpture? music? Or perhaps from the field of poetry? Or maybe science? And if you were asked to point out some of your creative pupils would your choice criteria for those pupils differ from the previous ones? why?

"*Creativity is the marvel of the human mind*", announced Margaret A. Boden (2004, p. 1) in her book "The creative mind: Myths and mechanisms". Furthermore, Boden acknowledges that "*Human creativity is something of a mystery, not to say a paradox. One new idea may be creative, while another is merely new. What's the difference? And how is creativity possible? Creative ideas are unpredictable*" (p. 1).

Although we share Boden's view of human creativity as something of a mystery, a marvel of the human mind, we believe it is not merely about "ideas", or some kind of "a product". There is also the creative process, the creative person and the creative environment (for further details see Chapter 1 of this book). This indeed raises many questions related to identifying "something" as creative, distinguishing it from a "regular" or a "common" occurrence.

In the last two decades, there has been a growing interest in the various issues related to creativity, acknowledging its central role in the success of individuals during their adult life and its contribution to the development of economics and society. At the same time, many aspects associated with the nurturing of creativity within the education system remained at the theoretical level of discussion, making it difficult to actualize theoretical ideas and suggestions into practical moves. The present book aims to bridge the existing gap between theory and practice. The editors of this book believe that the creativity of every individual can and should be fostered and that the education system needs to pay proper attention to the nurturing of creativity.

In order that teachers succeed in developing their pupils' creativity, at least two of the following prerequisites must be met: Teachers' acquaintance with the varied aspects related to creativity and adopting a belief about the importance of its nurturing among their pupils; familiarity with a variety of principles designed to help them develop learning environments that can support the fostering of student creativity.

Why is it essential to consider teacher knowledge and teacher beliefs? For the past four decades, issues concerning teacher knowledge and teacher beliefs have been extensively discussed in the research literature. In what follows, we briefly address these issues.

In-depth discussion of issues related to teacher knowledge began with the publication of the seminal paper by Lee Shulman's (1986), *Those who understand: Knowledge growth in teaching*. In this paper, Shulman proposed a distinction between three types of knowledge: subject matter content knowledge, pedagogical content knowledge, and curricular knowledge. Based on Shulman's conceptual framework, the field of teacher knowledge has evolved significantly, as has the context of mathematics teacher knowledge. Ball, Thames, and Phelps (2008), for example, developed the Mathematics Teacher's Specialisednowledge (MTSK) model, on which Carrillo-Yañez, Climent, Montes, Contreras, Flores-Medrano, Escudero-Ávila, et al. (2018) elaborated. All types of knowledge included in these models have direct implications for teachers' priorities, their teaching methods, the characteristics of the learning environments they generate, and more. These direct implications concern the development of pupils' mathematical thinking and attainments. The same is true for teachers' beliefs. Generally, beliefs affect individuals' interactions with the world and, in turn, they have an impact on their goals, perceptions, and behaviors. This is also relevant to teachers' interaction with their pupils in the school context, in particular, the nature of their instruction routines and educational goals (Richardson, 1996; Thompson 1992). Similar to findings regarding teacher knowledge, teachers' beliefs also have implications for student knowledge and achievements. Such results were extensively confirmed in the field of mathematics teaching and learning as well (e.g. Voss, Kleickmann, Kunter, & Hachfeld, 2013).

Given the above, this book suggests several ideas for designing learning environments aimed at supporting the development of pupils' mathematical creativity, as can be seen in the various chapters. The choice to illustrate the expression of these ideas with geometrical examples, stems from the fact that, among the content areas taught within school mathematics, geometry is perceived by teachers as the most appropriate for developing pupils' mathematical creativity, as well as an area that allows teachers to express creative approaches to teaching (Levav-Waynberg & Leikin, 2012).

On geometry and its teaching

Geometry is considered one of the areas of mathematics that has challenged humanity over the years. Moreover, geometry studies constitute an inexhaustible source of ideas as well as an instrument for developing logical and creative thinking. Today, geometry learning starts at pre-school and continues throughout all the years at primary, junior high and high school, as well as in higher education institutions and teacher education colleges. Furthermore, basic geometric terms are learned and built already during the first years of nursery school, constituting the foundations for understanding this field and the attitude towards it in the course of all one's years of learning.

Thales (624–547 B.C.), the Greek philosopher and mathematician, was one of the forefathers of geometry, formulating theorems and logical arguments. He was followed by Euclid at the end of the 4th century B.C. with his book *Elements*. Geometry, then, is a discipline that has been built in the course of more than 2500 years and many consider it as the masterpiece of human creation. Indeed, geometry constitutes an infrastructure for science and technology, viewed today as leading fields, and it is frequently implemented also in everyday life.

The National Council of Teachers of Mathematics (NCTM), which defined standards for mathematics teaching, indicates among others what pupils at different ages, from nursery school and up to the 12th grade, need to know in the field of geometry. These standards allow teachers to adapt the learning materials to the pupils' age and include in it all the competencies required for developing geometric thinking. The Standards document points out that these teaching plans should be considered by teachers who teach so that the pupils know how to "*analyze properties and determine attributes of two and three dimensional objects*" as well as "*develop mathematical arguments about geometric relationships*" (NCTM, 2000, p. 96). Hence, the main emphases in the teaching of geometry are designed to develop a visual perception in both plane and space. Moreover, the objective of teaching this discipline is to develop the skills and ways of thinking typical of geometry, among them logical thinking, raising hypotheses, reasoning and presenting arguments, drawing conclusions, identifying generalizations and so on.

As a result of its complexity, geometry is considered one of the more difficult branches of mathematics both for teaching and learning. Even the 21st century, that is so technology-intensive and highly advanced, has not essentially changed this situation. Furthermore, in many cases, geometry is perceived as a 'filter', since the high requirements inherent in it, facilitate examining the suitability of learners to higher education studies. Furthermore, international tests that serve as a measure for the level of study of each country compared to other countries include many geometric questions, thus increasing the prestige of the discipline. These characteristics of geometry are often the source of student anxieties and the belief that only virtuous individuals are capable of understanding it (Saglam, Turker, & Umay, 2011).

Several theories are associated with the development of pupils' geometric thinking. One of them, and perhaps the most familiar one, is the Van Hiele theory. This theory advocates that the development of and progress in geometric thinking does not depend on the pupils' age but rather on their stage of the learning. A series of studies of this topic (e.g. Clements & Battista, 1992; Koester, 2003; Pandiscio & Knight, 2011; Patkin & Barkai, 2014) found that difficulties in the identification of geometric figures and acquaintance with their properties, are manifested already among young pupils, as well as among some of the adult learners. Moreover, they are encountered by pre-service and in-service teachers. The problem resides in the fact that even experiencing within the framework of teacher education programs does not considerably enhance the knowledge, as it is measured through the Van Hiele theory (Pandiscio & Knight, 2011). Therefore, it is highly important to choose the type of activities in which pupils engage, emphasizing the full compliance between the pupils' level of understanding and the tasks with which they have to cope (Crowley, 1987).

In addition to the cognitive aspects mentioned above, geometry learning involves the use of varied visual competences. Visual competence constitutes also an important component in examining the pupils' Van Hiele level, already at the first stage (Van Hiele, 1999). Pupils with low visual competence, encounter difficulties while learning geometry. Consequently, one of the goals of geometry teaching is developing the pupils' visual capability and, as a result, their ability to solve problems. Visual competences can be developed, *inter alia*, by underscoring the applications and uses of geometry (Levenberg & Patkin, 2014). This issue is extensively referred to in the Standard document of the National Council of Teachers of Mathematics (NCTM, 2000). Geometry exists around. Geometric figures are part of the nature since time immemorial, where people's creations being only their imitation. When teachers use visual representations of all types, e.g. pictures and movies, exposing the pupils to geometry in their environment, this could serve as a bridge between the concrete and the abstract. Together with the encouragement and recommendation to apply visual representations, it is important to emphasize that every object around us can serve as an illustration of the learned material. Even a football or an orange can mark the beginning of a fascinating lesson. It all depends on the degree of imagination and creativity of teachers wishing to evoke interest and improve the attainments of their pupils in understanding the subject of geometry.

To conclude, the teaching and learning of geometry involve quite a few challenges. We believe that implementing the ideas presented in this book would help both teachers and pupils overcome some of the difficulties while developing creative mathematical thinking. Finally, we wish to point out that, although the book chapters deal with geometry, readers are encouraged to adapt the approaches to other areas of mathematics as well.

About this book

The book comprises three sections.

The first section was written by Atara Shriki. It consists of two chapters that expose the readers to a wide variety of aspects related to the term 'creativity' and its various expressions as well as to issues associated with the nurturing of mathematical creativity within the school framework. The first chapter in this section, *The Many Faces of Creativity*, presents several milestones in the development of the term 'creativity', starting from the Ancient Age and until the Modern Era. The chapter discusses various approaches to the investigation of creativity: psychometric, cognitive and social-personal. It also engages in several key issues in this context included in the empirical literature: general creativity versus content-dependent creativity; creativity from the aspect of the product and from the aspect of the process; and absolute creativity versus relative creativity. The second chapter, *Nurturing Mathematical Creativity at School*, presents some of the dilemmas and challenges related to the nurturing of mathematical creativity within the framework of the education system. In particular, it discusses questions associated with teachers' ability to develop the mathematical creativity of every learner, as well as the components of the learning environments designed for that purpose. The chapter distinguishes between 'creative teaching' (integration of interesting and activity-oriented teaching methods) and 'teaching for creativity' (identifying pupils' creativity and nurturing it). The chapter encompasses varied reasons why many teachers avoid nurturing the mathematical creativity of their pupils. In order to assist teachers in developing appropriate learning environments, the chapter suggests grounding the teaching and learning in approaches related to solving problems in various ways and in problem posing, which are two of the key features of the creative mathematical practice. Thus, these approaches can be easily integrated within the compulsory curricula.

The second section comprises five chapters (3–7), that depict creative approaches to the teaching of geometry.

Chapter 3 was written by Ilana Levenberg and Dorit Patkin. This chapter describes an inductive process of calculating the sum of internal angles and the sum of external angles of convex polygons by means of paper and scissors. Paper and scissors are instruments that facilitate the illustration of the terms and are also a scaffold for building the inductive process, aiming to attain generalization and better comprehension of the terms. The learning environment offers the pupils an opportunity to conduct independent inquiry activities, raise questions, discover properties, generalize and provide arguments.

Chapter 4 was written by Ronit Bassan-Cincinatus. It engages in the development and enhancement of areas and perimeters calculation skills while employing cognitive abilities at different levels. The chapter illustrates how the combination of two familiar basic figures – square and circle – results in the formation of complex figures that can be decomposed in different ways. Thus, areas and perimeters can be calculated in a creative manner. The activities presented in this chapter constitute an example of an approach to developing mathematical creativity by problems that, apparently, have no immediate solution; problems that have several solutions; and problems that allow various approaches to problem solving.

Chapter 5 was written by Ilana Levenberg. The chapter encompasses activities designed to promote graphical literacy and nurture creative thinking. This is achieved by developing logical thinking and in-depth comprehension of the meaning of information presented by a coordinate system. The tasks included in this chapter stimulate pupils to cope in various ways with unusual, non-algorithmic problems while delving into completing missing information and asking questions.

Chapter 6 was written by Dorit Patkin. The chapter engages in proving and refuting claims. These activities are important for building geometric understanding since coping with them constitutes a valuable instrument for the promotion of creative thinking. Pre-service teachers had to prove and refute claims in order to examine their ability to identify whether claims are true or false and present their reasoning. Such activities enhance the awareness of accurate formulations and the meticulous use of accurate mathematical language. This can help pre-service and in-service teachers to promote their pupils' mathematical and deductive thinking as well as their argumentation ability.

Chapter 7 was written by Atara Shriki. It presents the place of problem posing processes in the development of mathematical creativity, illustrating how this approach can be easily implemented by applying the 'what if not?' strategy in various situations. Illustration of the process is based on the problem presented by Ilana Levenberg and Dorit Patkin in Chapter 3 of this book. It indicates the way problems at various complexity levels can be posed as well as the ability of all the teachers to develop by themselves learning materials in this spirit.

The third section engages in geometry and the creative brain.

Chapter 8 was written by Eli Appelboim. It leads us by measured steps from Euclidean geometry to non-Euclidean geometry and its innovations in the context of surfaces with changing curvature and up to signals and pictures processing and what lies between them. The chapter transmits the message that it is essential to expand the learning of geometry and the inculcation of terms from the field of non-Euclidean geometry already at school. The purpose is to prepare pupils to be integrated into the forefront of the science and technology industry, in which mathematics in general and geometry in particular play a major role. Moreover, this is designed to inculcate an original thinking ability and a broad picture of the world among all pupils. The examples included in the chapter represent innovative and exceptional outcomes and, thus, they illustrate products of creative thinking.

Chapter 9 was written by Ilya Sinitsky and Moshe Stupel. The chapter presents the concept of variance and invariance as a geometric and didactic instrument. The writers present unusual evidence of geometric occurrences through a unifying lens of invariance. Moreover, they describe optional changes in geometric objects, while affecting their properties. This creative approach to the exploration of theorems and generalizations in geometry offers a unique instrument for proving claims and solving problems in unusual ways.

Chapter 10 was written by Atara Shriki. The chapter presents an approach to creativity assessment and the development thereof in the context of problem posing, based on psychometric indices. The psychometric indices are fluency, flexibility, originality, and organization (or elaboration), and relate to both absolute creativity and relative creativity. This approach can be easily implemented, allowing teachers to monitor changes transpiring among their pupils. Furthermore, it offers pupils a simple means of developing their ability to self-assess the development of their creativity. The chapter ends with a reservation regarding the exclusive reliance on the assessment of creativity by psychometric indices.

We conclude the book with an Epilog that comprises seven challenging geometric problems that call for implementing creative thinking. Then we present some information about the editors and the writes of the book.

We wish our readers much pleasure in reading the book and hope that it will assist them in the implementation of learning environments that support the development of their pupils' mathematical creativity.

Dorit Patkin, Atara Shriki, and Ilana Levenberg

October 2020

References

1. Ball, D. L., Thames, M. H., & Phelps, G. (2008). Content knowledge for teaching: What makes it special? *Journal of Teacher Education, 59*(5), 389–407.
2. Boden, M. A. (2004). *The creative mind: Myths and mechanisms* (2nd ed.). London: Routledge.
3. Carrillo-Yañez, J., Climent, N., Montes, M., Contreras, L. C., Flores-Medrano, E., Escudero-Ávila, D., Carillo & Muñoz-Catalán, M. C. (2018). The mathematics teacher's specialized knowledge (MTSK) model. *Research in Mathematics Education, 20*(3), 236–253.
4. Clements, D. H., & Battista, M. T. (1992). Geometry and spatial reasoning. *In* G. Douglas, (Ed.), *Handbook of research on mathematics teaching and learning* (pp. 420–464). New York. NY: MacMillan.
5. Crowley, M. L. (1987). Van Hiele model of the development of geometric thought. In M. Montgomery Lindquist, & A. P. Shulte (Eds.), *Learning and teaching geometry*, K-12, 1987 Yearbook of the National Council of Teachers of Mathematics (NCTM) (pp. 1–16). Reston, VA: NCTM.
6. Koester, B. A. (2003). Prisms and pyramids: Constructing three-dimensional models to build understanding. *Teaching Children Mathematics*, *9*(8), 436–442.
7. Levav-Waynberg, A., & Leikin, R. (2012). Using multiple solution tasks for the evaluation of pupils' problem-solving performance in geometry. *Canadian Journal of Science, Mathematics and Technology Education, 12*(4), 313–333.
8. Levenberg, I., & Patkin, D. (2014). Promoting meaningful learning - Studying mathematics with traffic signs. *International Journal of Learning and Development (IJLD), 4*(2), 1–8.
9. National Council of Teachers of Mathematics (NCTM). (2000). *Principles and standards for school mathematics*. Reston, VA: NCTM, 2000.
10. Pandiscio, E. A., & Knight, K. C. (2011). An investigation into the Van Hiele levels of understanding geometry of preservice mathematics teachers. *Journals of Research in Education, 20*(1), 45–52.
11. Patkin, D., & Barkai, R. (2014). Geometric thinking levels of pre- and in-service mathematics teachers at various stages of their education. *Educational Research Journal, 29*(1&2), 1–26.

12. Richardson, V. (1996). The role of attitudes and beliefs in learning to teach. In: J. Sikula, T. Buttery, & E. Guyton (eds.), *Handbook of research on teacher education*, 2nd ed. (pp. 102–106). New York, NY: MacMillan.
13. Saglam, Y., Turker, B., & Umay, A. (2011). Geometry anxiety scale for secondary school students. *Procedia Social and Behavioral Sciences, 15*, 996–970.
14. Shulman, L. S. (1986). Those who understand: Knowledge growth in teaching. *Educational Researcher, 15*(2), 4–14.
15. Thompson, A. G. (1992). Teachers' beliefs and conceptions: A synthesis of the research. In D. A. Grows (ed.), *Handbook of research on mathematics teaching and learning* (pp. 127–146). New York, NY: MacMillan.
16. Van Hiele, P. M. (1999). Developing geometric thinking through activities that begin with play. *Teaching Children Mathematics*, *5*(6), 310–316.
17. Voss, T., Kleickmann, T., Kunter, M., & Hachfeld, A. (2013). Mathematics teachers' beliefs. In M. Kunter, J. Baumert, W. Blum, U. Klusmann, S. Krauss, & M. Neubrand (Eds.), *Cognitive activation in the mathematics classroom and professional competence of teachers: Results from the COACTIV Project* (pp. 249–271). Boston, MA: Springer.

Section I

The World of Creativity

Photo by S. Gazit

https://doi.org/10.1142/9789811218750_0001

Chapter 1

The Many Faces of Creativity

Atara Shriki

Introduction

In his book *A Whole New Mind*, Daniel Pink (2005) argues that we are on the verge of a new age, whereby creativity, intuition and the ability to link objects and events that apparently are unrelated, turning them into something new and different, will constitute the most significant skills.

For over a hundred years researchers have explored creativity, its characteristics, roots, and development (Plucker, Beghetto, & Dow, 2004), while attempting to define these issues and ground them in a wide empirical context. These studies yielded a variety of insights about and explanations for the nature of creativity, its origins, assessment of its development over the years (Andiliou & Murphy, 2010) and its manifestation among people (Silver, 1997).

Review of the empirical literature in the field of creativity, in general, and mathematical creativity, in particular, reveals more than one hundred contemporary definitions of creativity as well as numerous interpretations of its essence and meaning (Mann, 2006). Due to the complexity of this area, a considerable part of these definitions and interpretations are rather vague (Sriraman, 2005) and to date, none of the definitions has been universally accepted. Moreover, creativity has been investigated throughout the years from at least four perspectives: the creative process, the creative person, the creative environment and the creative product (Henry, 2009). These perspectives yielded a variety of approaches to the assessment of creativity, among them a psychometric approach (e.g. Guilford, 1967; Torrance, 1974), a cognitive approach (e.g. Sternberg & Davidson, 2005), and a personal-social approach (e.g. Sternberg & Lubart, 1996; Wallas, 1926).

In light of the above, it is only natural that Henry (2009) sums up that creativity "*seems to be one of those words that, although commonly used, is not easy to define. We may use the term regularly, but can struggle if asked to put into words specifically to what we are referring*" (p. 200).

This chapter presents several of the main milestones in the development of the concept of "creativity". It distinguishes between general creativity and specific creativity, between absolute creativity and relative creativity, and between the creative process and creative product.

Milestones in the Development of the Concept "Creativity"

Creativity in the Ancient World, the Middle Ages and the Renaissance Period

In his book that was translated from Polish, Tatarkiewicz (1980) describes the way in which the perception of "creativity" has changed and developed, starting in the period of Ancient Greece. In Ancient Greece, there was no concept that matched the meaning of "creation" or "creator". The nearest concept was the verb "to do" and it mainly related to poetry. Poetry was not perceived by the Greeks as bound by any laws and, hence, poets were considered the closest to "creative people". In other words, the Greeks viewed "creativity" and "freedom of doing" as two equivalent terms. However, this was not the case in the context of art and music, for example, areas that today are considered as areas that express creativity. The Greeks perceived art as an imitation of reality rather than as the creation of reality. They maintained that painters' role was discovering the laws of nature and not inventing them. Melodies were also considered as something "dictated", particularly those played in ceremonies and performances designed to entertain the audience. In fact, composers had no freedom of doing or creating.

It is noteworthy that in a certain sense, this perception is similar to one of the common controversies in the field of mathematics, namely: Is mathematics a discovery or an invention? Some researchers argue that every mathematical object exists, whether it is discovered or not. Hence, mathematicians do not invent or generate mathematics but only discover it. Conversely, others maintain that mathematics is not a discovery but rather an invention of the human brain or the creation of new knowledge (Davis & Hersh, 1981; Ernest, 1991).

The Romans challenged the perceptions of the Greeks. According to them, poets, as well as painters and sculptors, were imaginative creators, having the freedom to dare and do whatever they pleased. The term "imagination" was new to the world of creation and in Latin, a unique term for "creation" was coined. In the Christian Period that followed, the concept "creation" underwent a transformation: God's action of creating something out of nothing. People's creations became once again "doing". Later, during the Middle Ages, poetry and painting became "art", a concept that was perceived as essentially different from "creation".

The Renaissance Period entailed a renewed flourishing of the concept "creativity". During that period, high importance was attributed to the meaning of independence, freedom, and creativity. Poetry, painting, and sculpturing became a "creation" again, and so did architectural buildings. Contrary to the Greeks' perception, Renaissance people viewed art as something that transcends nature rather than its imitation. They perceived artists and poets as designers of a new world and composers as creators of new music and new melodies.

Creativity in the Modern Period

In the 18th century, theories engaging in creativity started developing, and mainly the concept "imagination" was connected to them. However, not everybody linked the concept "imagination" to creation. Some regarded creativity as the ability to remember shapes and contents, connect them in some way and change their size. Some viewed "creation" as a mysterious doing, something that clashed with the psychological conceptions of those days. Others, on the other hand, still embraced the opinion that "creation" belongs to God. In the 19th century, no one had any doubt that art was a creation. Furthermore, many believed that art was the only expression of creativity. Only in the 20th century, people began discussing creativity also in the context of sciences (Tatarkiewicz, 1980).

Studies of intelligence in general and genius in particular, conducted in the middle of the 19th century and beginning of the 20th century, constitute the foundation for the development of systematic research in the field of creativity. The first attempts of engaging scientifically in issues related to intelligence and the characteristics of genius are attributed to the English psychologist Sir Francis Galton (1822–1911), often considered as "the first cognitive psychologist". Galton started his work on these topics by a self-inquiry of his brain activity. In 1879 he wrote down some of his findings and explained the motivation for his work by his wish to understand how a sequence of associations turns from something vague that crosses the threshold of awareness into something conscious (Dacey, 1999). Following Galton, additional attempts were made to systematically explore various aspects of intelligence. The French psychologist Alfred Binet and his disciple Theodore Simon, contemporaries of Galton, developed a series of intelligence tests that were first published in 1908 (Wakefield, 1991). The improved version of the Binet-Simon test consisted of 30 assignments listed by increasing order of difficulty, the first assignments being those that every person can perform.

During the 1950s, people began relating to the issue of creativity also in a scientific way. Researchers such as Joy Paul Guilford, Alex Osborn, Genrikh Altshuller and Edward de Bono are considered today as pioneers in the field of the study of creativity. At the same time, researchers started showing interest also in the advancement of creativity in an educational context (Craft, 2001a), believing that creativity can motivate economic and social changes (Lin, 2011).

Guilford was the former president of the American Association of Psychologists and one of the precursors of the psychometric approach to creativity. He mainly engaged in the development of psychometric methods of measuring human intelligence, while distinguishing between two types of response to a given problem: convergent thinking and divergent thinking (Guilford, 1967). He defined convergent thinking as the ability to give one "correct" answer to standard problems (like most of the problems presented in the school context). Such thinking is oriented at finding the most "suitable" answer, putting an emphasis on memorization, speed, accuracy, and implementation of familiar techniques. On the other hand, divergent thinking emerges spontaneously, as a flow of ideas with unexpected connections. For example, multiple solutions to the same problem are the product of divergent thinking. Guilford associated divergent thinking with creativity. According to him, following the presentation of diversified ideas, the function of divergent thinking is to organize them logically, aiming to make decisions. Guilford's ideas constituted the basis for investigating the characteristics of creative people. Studies (e.g. Chamorro-Premuzid, 2006; McCrae, 1987) illustrated that there are two prominent personality characteristics related to divergent thinking, i.e. openness and extroversion. Nevertheless, no prominent personality characteristics were found in relation to people who tend to apply convergent thinking.

Another milestone in the study of creativity was conceived by Torrance (1974), who developed the Torrance tests of creative thinking, based on the ideas of Guilford (1967) in the context of creativity. Torrance's tests, many of which are prevalent also today, engage in problem solving. The examinees' scores relate to four indices (Torrance, 1974): fluency — the total number of interpretations or relevant meaningful ideas presented as a response to external stimulus or an associative sequence based on the use of universal knowledge; flexibility — the number of different of relevant response categories or the ability to change ideas; originality — the extent to which the responses are rare or the uniqueness of the product; elaboration — the number of details included in the responses or the ability to generalize ideas. Among the indices encompassed in Torrance's definition, originality (or novelty) was acknowledged as the most consensual aspect for describing creativity, since creativity is usually perceived as creating ideas, approaches and original products (Leikin, 2009; Shriki, 2010).

During the 1960s, Torrance and his colleagues from the University of Minnesota explored the creativity of thousands of primary school pupils (obviously, it takes too long to mention them). With time, Torrance expanded his tests, adding criteria such as emotional expressiveness, storytelling articulateness, movement or action, synthesis of incomplete figures, unusual visualization, extending or breaking boundaries, the richness of imagery, and more (Ball & Torrance, 1984; Torrance, Ball, & Safter, 1992). Torrance's tests, as well as tests developed over the years on this basis, are widely used today in the research and assessment of creativity.

Unlike the psychometric approach, the cognitive approach to the study of creativity focused on the cognitive and mental processes involved in creativity, including the use of various representations, creation of mental connections between objects that apparently are not inter-related and the ability to solve problems from various areas (Sternberg & Davidson, 2005). In fact, the cognitive approach attempts to decode the processes performed mentally on the entirety of knowledge stored in the long-term memory as well as on the way in which parts of information are related, resulting in a creative product. The Triarchic Theory of Intelligence conceived by Sternberg (1985), constituted a breakthrough by being the first theory that advocated against the psychometric approaches to measuring intelligence while putting an emphasis on a cognitive approach. Sternberg linked between the way consciousness functions and information is processed, and the three-component types: meta-components, performance components and knowledge acquisition components. According to him, meta-components are processes used for solving problems and making decisions. These are the key processes of human consciousness since they direct people's actions. Performance components are the processes that execute what has been created through the meta-components. These are the basic processes that enable people to perform tasks. The knowledge acquisition components are used for acquiring new knowledge, facilitating the choice of information out of irrelevant information or the combination of knowledge parts. Sternberg (1997) extended the Triarchic Theory of Intelligence to a sub-theory engaging in gifted learners, arguing that they apply these components more efficiently than others. More specifically, he examined the relationship between the nature of the task performance and the extent of it being known to the performer while dividing the role of previous experience into two parts: innovativeness and automation. A new situation is a situation that people have not yet experienced. People capable of performing new tasks can find new ways of dealing with them, ways that most people would not notice. A process performed automatically is a process that has already been performed several times and can be executed without particular thinking. Once the process becomes automatic, it can be performed parallel to other processes. Cianciolo and Sternberg (2004) maintain that success in any area of our life depends considerably on individuals' ability to exploit their analytical, creative and practical skills.

Another approach to the study of creativity is the personal-social approach. This approach underscores the emotional aspects of creativity and social-cultural features underpinning it (Sternberg & Lubart, 1995). According to this approach, creativity is a consequence of interaction with the environment (Sawyer, 1995), and the interaction between personality traits and environmental elements greatly affects the actualization of creative ability. In recent years, Glăveanu (2010) enhanced the social approach to the investigation of creativity, positioning it in the wider context of culture. Based on previous studies of this field, he defined the meaning of "creativity from a cultural perspective" as "*a complex socio-cultural-psychological process that, through working with 'culturally-*

impregnated' materials within an intersubjective space, leads to the generation of artifacts that are evaluated as new and significant by one or more persons or communities at a given time" (p. 89). This approach forms a framework by means of which the relation between the product, the creator and the community can be examined. This is done by relating to cultural aspects, existing norms and the way by which the community builds the creator's self-identity. In other words, Glăveanu (2010) maintains that one cannot relate to creativity as to something objective. It can only be seen in the context of a community that examines the creative product based on its previous knowledge and cultural repertoire, through a dialog between the community and the creator.

And what are the new trends in the study of creativity over the last two decades? Nowadays, we witness new trends associated with the investigation of creativity. While research of creativity is expanding and developing according to the various approaches mentioned above, at the same time, with the development of technology, researchers in the field of brain sciences have also started exploring this topic. One of the leads relates to the functional Magnetic Resonance Imaging (fMRI) that provides indices for neural brain activity according to changes in the blood flow (e.g. Abraham, Thybusch, Pieritz, & Hermann, 2014; Huettel, Song, & McCarthy, 2009). Eric Kandel, who in 2000 was one of the Nobel Prize winners in the field of physiology and medicine, was one of the precursors of brain research by means of fMRI. Kandal (2006) believes that the clearer the relation between brain activity and creativity, the easier the understanding of the situations that enhance the development of creativity. Another technology that was recruited to the study of creativity is the eye-tracking technology (e.g. Schindler & Lilienthal, 2017; Schindler, Lilienthal, Chadalavada, & Ögren, 2016). These research areas are still at the very beginning. However, they undoubtedly have the potential to reinforce and enhance the way we understand the meaning of creativity and its diversified expressions.

General Creativity versus Domain-Specific Creativity

Researchers do not concur about the question of whether creativity is domain-specific or perhaps it is a general ability (Craft, 2001a). One of the main reasons they fail to draw a sweeping conclusion stems from differences between the research instruments, research population and disciplines in various studies. Some researchers view creativity as a general ability (e.g. Beghetto & Kaufman, 2009) while others consider it is an ability resulting from individuals' interaction with a specific domain (e.g. Csikszentmihalyi, 1999). General creativity is related to the ability to connect the solution of problems in one field to their solution in another field or to the ability to implements patterns of problem solutions from a certain field in order to solve problems in another field. Domain-specific creativity, however, is associated with the ability to apply the logical and deductive rules of a particular discipline (Haylock, 1987; Piirto, 1999).

The literature review of Sriraman (2005) illustrates that those who perceive creativity as a general ability, maintain that creativity is controlled by the same cognitive processes, regardless of the discipline. According to them, studies indicate that creators who are prominent in their field, point out the same stages of creation preparation, incubation, illumination and verification of their ideas (e.g. Wallas, 1926). According to Sriraman (2005), the difference between a creative and non-creative process resides only in the scope of activities. Others (e.g. Lucas, 2001), believe that creative skills are manifested in daily life regardless of the discipline. Moreover, Welling (2007) claims that in a certain sense every person can be considered creative since no situation is similar to its predecessor. Young children, for example, develop by discovering something new every day. Hence, these skills should be exploited in each of the fields studied at school and in every aspect associated with our daily life. The literature review carried out by Neumann (2007) indicates that the different disciplines share common characteristics also in the aspect of the personal traits of the great creators. For example, high appreciation of aesthetics, high degree of consciousness, extensive areas of interest, inquisitiveness and attraction to complexity, strong intuition, ability to cope with contradictory information, agreeable and pleasant conduct or, conversely, neurotic and even anxious behavior.

The proponents of the opposite approach argue that creativity is domain-specific. They ground their arguments in the works of creators who are prominent in their field. In their opinion, the fact that very few creators are responsible for a large number of important creations indicates that general, daily creativity does not require the same skills manifested in creativity in a specific area. Thus, for example, prominent creators in the field of science need about 20 years in order to reach the peak of their work. Consequently, the creative process these people are undergoing cannot be compared to the process undergone by people who solve simply, non-algorithmic problems, the solution of which requires an illumination that can be attained within a few minutes (Sriraman, 2005). Moreover, other studies (e.g. Feldman & Benjamin, 2006; Kaufman & Beghetto, 2009; Milgram & Livne, 2005) illustrate that a creative expression requires knowledge in a specific area of knowledge.

Perception of Creativity from the Product and Process Aspects

Researchers in the field of cognitive psychology distinguished between product-oriented creativity and process-oriented creativity while relating to the various aspects of an invention that is innovative and valuable (James, Lederman, & Vagt-Traore, 2004; Smith, 2005). Product-oriented creativity is based on the assumption that at the heart of creativity stands a product or an outcome that is both innovative and useful. In this context, "innovative" relates to an original work and "useful" relates to the fact that the product satisfies a certain need (Sternberg & Lubart, 1995). Process-oriented creativity focuses on mental processes that are connected to the generation of new ideas, solutions to problems and self-fulfillment (Fryer, 1996). Other researchers related to creativity by the terms "big-C" and "little-c" (Craft, 2001b).

The first term refers to the extensive effect that creation has on society while the second refers to the relevance of creativity to our day-by-day life in the sense that every person has the potential to be creative and solve problems in daily life.

Creativity from the product aspect

Many researchers in the field of creativity relate to its expression mainly from the aspect of the final product, particularly a product that is new, unexpected and different from any other product (Sternberg & Lubart, 1995, 2000), original and unusual (Weisberg, 1993), useful (Sternberg & Lubart, 1995, 2000) and provides an exceptional or innovative response to a given problem (Torrance, 1974). Yet, according to Shye and Yuhas (2004), innovativeness in itself has no value, unless it provides a suitable response to a challenging situation. A creative solution should be assessed, like any other solution, not only because of its innovativeness but also due to its extent of appropriateness and suitability to the challenge one copes with. Reference to the issue of "appropriateness" has been examined from several aspects, among them the extent to which the solution is efficient, useful, satisfies a social need (Plucker *et al.*, 2004) and provides the correct answer to the problem (Sternberg & Lubart, 1995). In other words, the meaning of creativity as yielding a product that is not merely innovative but also appropriate is that the product should be useful for the purpose for which it was designed. People who "invent" wild ideas that have no practical application, might be sometimes perceived as "unbalanced" rather than creative (Simonton, 1999). As an example, Simonton quotes Thomas Edison's invention of the light bulb. It is perhaps not an exceptional creation in its complexity but its value will be remembered for many years because of its degree of usefulness. This idea leads to the conclusion that some products are considered more creative than others. That is, certain creativity is more impressive than others since the product is more revolutionary than others or is entirely new. For example, Rembrandt painted remarkable paintings but his painting style was in line with the prevalent tradition of those days. Conversely, Van Gogh paved the way to a new style of painting; Mozart is considered today as one of the most gifted musicians ever who took the musical tradition of his days to new peaks. However, Arnold Schönberg is viewed as the composer who revolutionized music. Faraday contributed a lot to the understanding of the electric particle and magnetism but after Einstein and Bohr, physics was entirely changed. Hence, when we relate to the product from the aspect of creativity, we have to distinguish between “important” and “revolutionary”. While most of the new discoveries and theories transpire within the existing methodological framework, only few exceed and transcend its boundaries (Kuhn, 1962).

We should point out that the psychometric approach mentioned above, and particularly the tests conceived by Guilford (1967) and Torrance (1974) mainly measure the creativity degree of the product. However, as mentioned, Torrance added measures to the initial tests he had developed, referring also to personality and emotional criteria.

Creativity from the process aspect

The above-mentioned cognitive approach related to creativity mainly from the aspect of the process, although many researchers argue that it is difficult to accurately characterize the cognitive process underpinning creative thinking. Welling (2007) indicates that Plato (cited in Jowett, 1937) mentions the “creativity paradox” according to which people cannot produce ideas that are beyond what is in their brain. Plato’s solution to this paradox is that all new knowledge is, in fact, a reminiscence, even from previous lives. In spite of Plato’s mystic perception, many studies of creativity focus on the attempt to comprehend the process characteristics and stages. The literature review of Welling (2007) illustrates a controversy regarding the characteristics of the creative process. This controversy is associated with the question of whether this concerns a gradual, continuous process that facilitates finding the creative solution or perhaps this is a single process whereby the solution surfaces in a sudden way. Proponents of the gradual process present as evidence occurrences such as "sense of knowledge" and "feelings of warmth", as well as the fact that creative people usually need many years in order to obtain creative solutions. This continuous cognitive process is partly conscious and partly unconscious. Conversely, advocates of the process as a "single sudden leap" claim that they have failed to find evidence for the "feelings of warmth towards the solution". They maintain that these are distinct stages that creative people undergo, such as release from fixation, reconstruction of the problem and illumination.

Unfortunately, most mathematicians do not demonstrate interest in analyzing their thinking processes that lead to mathematical creation (Ervynck, 1991) and only few attempts have been made to document these processes. Hence, the contribution of the prolific French mathematician Jacques Hadamard is prominent in this respect. Hadamard (1945) interviewed a hundred outstanding mathematicians and physicists, trying to describe their thinking process, as well as his own, combining the findings in his book entitled "*The psychology of invention in the mathematical field*". - His main findings were that the prominent creators' thinking process transpired mostly without words but rather by mental images that emerged suddenly and spontaneously, until becoming an "illumination".

Hadamard (1945) was greatly influenced by Gestalt psychology, particularly by the model of the English psychologist Graham Wallas. Wallas (1926) described creativity as a 5-stage process: Preparation (focusing on the problem and investigating its dimensions in order to get an insight into the problem); Incubation (putting the problem aside for a while, internalizing the problem into the subconscious without an external manifestation); Intimation (linking Incubation and Illumination, "sensing" that the solution is imminent); Illumination (the solution suddenly appears, as the creative idea passes from the subconscious to the conscious); and Verification (the idea is verified in a conscious level, possibly extended through utilization of the result). Hadamard (1945), like many others, related only to four out of the five stages, omitting the "Intimation" stage.

Many years after Hadamard (1945) had published his book, Liljedahl (2008) chose five out of the 33 questions in Hadamard's original interview. These questions were directly related to mathematical discovery and mathematical creation and he circulated them to 25 leading mathematicians. Liljedahl revealed 12 themes that emerged from the data, each of which could be attributed to one of the four stages included in the model of Hadamard (1945). While some of these themes supported the findings of Hadamard, others served to extend it. Liljedahl (2008) salient findings were: (i) During the preparation stage, the mathematicians attributed more importance to the transmission of mathematical knowledge through talking than through reading; (ii) Throughout the incubation stage, mathematicians de-emphasize the role of details, arguing that getting down to details at this stage disrupts the thinking process and causes digression from it; (iii) The participants' responses indicate that they view the illumination stage as involving intrinsic chance ("luck of coming up with an answer, of having the right combination of ideas join within your mind to produce a new understanding" (Liljedahl, 2008, p. 157), as well as extrinsic chance ("the luck associated with a chance reading of an article, a chance encounter, or some other chance encounter with a piece of mathematical knowledge, any of which contribute to the eventual resolution of the problem that one is working on" (Liljedahl, 2008, p. 158). In this regard, Liljedahl commented that the idea that mathematical discovery often relies on unpredictable occurrences of chance encounters (even while sleeping, driving or taking a shower) is utterly contradictory to the prevalent image of mathematics as a field subjected to logic and deductive reasoning.

It should be noted that both the model of Wallas (1926) and that of Hadamard (1945) do not actually define creativity itself; rather, they try describing the stages of the process involved in it. Hence, a large part of the creative process remains, in fact, vague since it is unclear what transpires during the incubation stage. Moreover, we do not know how the transition from the subconscious to the conscious occurs, leading to the moment of illumination. Another limitation of the models stems from the fact that they refer to the process of solving problems that have already been posed and, thus, it ignores the process that preceded the posing of the problem (Sriraman, 2009).

The model conceived by Ervynck (1991) relates partly to the way by which the problems are posed. He described three stages which comprise mathematical creativity: Stage 0 — a preliminary technical stage, consisting of a technical or practical application of a mathematical rules or procedure, without considering their theoretical foundations; Stage 1 — algorithmic activity consisting mainly of applying explicitly mathematical operations and techniques in a sequential manner; stage 2 — creative (conceptual, constructive) activity, during which "*a true mathematical creativity is likely to occur and act as a motive power in the development of a mathematical theory. A non-algorithmic decision is made in a manner which seems to signify a bifurcation of the underlying concept structure. Mathematical creativity is the ability to perform such steps. The decisions that have to be taken may be of a widely divergent nature and always involve a choice, such as a choice of a certain concept to be defined*" (Ervynck, 1991, p. 43). Note that this model also has limitations, since it does not characterize in detail stage 0 and stage 1, the stages during which the problems are posed.

Another attempt to describe the creative process was done by Welling (2007). He stipulated that creativity is a result of four cognitive actions: application, analogy, combination, and abstraction. Application is a cognitive process relating to the individuals' ability to use existing knowledge in order to cope with a certain situation or problem; analogy is the individuals' ability to relate to a term taken from a certain context and use it in another context; combination occurs when the individuals merge two or more terms for the purpose of generating a new idea; and abstraction is a cognitive action that depicts mental processes and the result thereof.

Early experiments in the field of brain sciences, aiming to describe the creative process (e.g. Martindale, 1989; Shriki & Bar-On, 1997), refer to theories that engage in attention. Martindale (1989) argued that focused concentration activates a small number of adjacent junctions in the cortex. These junctions attain a high level of activity, whereas the other junctions are not active at all or are active on a low level. When attention is split, a large number of links among nodes become active on an intermediate level. According to the model, the illumination stage occurs when attention is widely spread and numerous associative paths that interconnect the junctions become active. Consequently, enhanced vigilance, combined with motivation or over-exertion to concentrate and obtain a solution, might inhibit creative thinking. In many senses, this process is connected to the divergent and convergent thinking discussed by Guilford (1967). Divergent thinking is associated with intermediate activation of nodes in the cortex that are far from each other, whereas convergent thinking implies a high level of activation of nodes in the cortex that are close to each other (Shye & Yuhas, 2004). It would be interesting to examine these theories through the efforts exerted today in the field of brain sciences.

Absolute creativity versus relative creativity

The empirical literature engaging in creativity from the aspect of the product sheds light on the need to distinguish between absolute creativity and relative creativity. The reason is that reference to creativity from the viewpoint of an innovative and useful product restricts the ability to examine learners' products in the sense of "creative product". Hence, the question raised is: "What is the level of innovativeness that defines someone or something as creative?".

Most of the definitions that attribute to creativity the production of an innovative product are suitable only to researchers who are leading in their discipline (Chamberlin & Moon, 2005). Consequently, there is frequently a tendency to view mathematical creativity as an ability that can be attributed only to professional mathematicians (Sriraman, 2005). However, in the school context, pupils cannot be expected to produce a mathematical product that is new in an absolute way (Koichu & Andžāns, 2009). This raises the question: "Is the meaning of creativity only a discovery of an absolutely original and innovative result? Or is it perhaps an independent discovery of a mathematical result or mathematical strategy that were unfamiliar to the learners (although they were known to the mathematical community)?". Since the definition of creativity in terms of innovativeness and usefulness is irrelevant to the identification of creative thinking processes in the school context, some researchers tend to distinguish between creativity of professionals in the discipline and creativity on the school level (e.g. Leikin, 2009; Liljedahl & Sriraman, 2006; Shriki & Lavy, 2012; Sriraman, 2005). These researchers stipulate that even if a discovery might not be considered innovative in an absolute manner, then in the school context pupils' discovery should be considered creative when it is examined from a personal perspective. The outcomes produced by pupils have to be innovative for them (Beghetto & Kaufman, 2009). Thus, pupils' creativity is relative creativity since their ability to consolidate an idea or mathematical solution of a problem they have not learned before, or to create an original solution for a problem studied before, is examined against the pupils' background and, in comparison with other pupils.

Evidence of support for this idea can be found in the words of prominent mathematics like Jacques Hadamard (1945) and Georg Polya (1954). They argue that the only difference between a creative work of professional mathematicians and that of pupils is the level of the work. Every person acts according to his or her relative level and therefore we should acknowledge that pupils are also capable of producing a creative product.

These arguments naturally lead to the second part of the Introduction which explores the place of creativity in general and mathematical creativity in particular in the school context.

References

1. Abraham, A., Thybusch, K., Pieritz, K., & Hermann, C. (2014). Gender differences in creative thinking: behavioral and fMRI findings. *Brain Imaging and Behavior, 8*(1), 39–51.
2. Andiliou, A., & Murphy, K. P. (2010). Examining variations among researchers and teachers' conceptualizations of creativity: A review and synthesis of contemporary research. *Educational Research Review, 5,* 201–219.
3. Ball, O. E., & Torrance, E. P. (1984). *Torrance Tests of Creative Thinking: Streamlined scoring guide Figural A and B*. Bensenville, IL: Scholastic Testing Service.
4. Beghetto, R. A., & Kaufman, J. C. (2009). Do we all have multi-creative potential? *ZDM - The International Journal on Mathematics Education, 41,* 39–44.

5. Chamberlin, S. A., & Moon, S. M. (2005). Model-eliciting activities as tool to develop and identify creatively gifted mathematicians. *Journal of Secondary Gifted Education, 17*(1), 37–47.
6. Chamorro-Premuzic, T. (2006) Creativity versus conscientiousness: Which is a better predictor of student performance? *Applied Cognitive Psychology, 20,* 521–531.
7. Cianciolo, A. T., & Sternberg, R. J. (2004). *Intelligence: A brief history*. Malden, MA: Blackwell Publishing.
8. Craft, A. (2001a). *An analysis of research and literature on creativity in education.* Report prepared for the Qualifications and Curriculum Authority. Retrieved from http://www.euvonal.hu/images/creativity_report.pdf
9. Craft, A. (2001b). Little C creativity. In: A. Craft, B. Jeffrey, & M. Leibling (Eds.), *Creativity in education* (45–61). London: Continuum.
10. Csikszentmihalyi, M. (1999). Implications of a system perspective for the study of creativity. In: R. Sternberg (Ed.), *Handbook of creativity* (313–335). Cambridge, England: Cambridge University Press.
11. Dacey, J. (1999). Concept of creativity: A history. In: M. A. Runco & S. R. Ritzker (Eds.), *Encyclopedia of creativity, Vol. 1* (309–322). San Diego, CA: Academic Press.
12. Davis, P. J., & Hersh, R. (1981). *The mathematical experience*. New York: Houghton Mifflin.
13. Ernest, P. (1991). *The philosophy of mathematics education*. London: Routledge.
14. Ervynck, G. (1991). Mathematical creativity. In: D. Tall (Ed.), *Advanced mathematical thinking* (42–53). Dordrecht, Netherlands: Kluwer.
15. Feldman, D. H., & Benjamin, A. C. (2006). Creativity and education: An American retrospective. *Cambridge Journal of Education, 36,* 319–336.
16. Fryer, M. (1996). *Creative teaching and learning*. London: Paul Chapman Publishing Ltd.
17. Glăveanu, V. (2010). Paradigms in the study of creativity: Introducing the perspective of cultural psychology. *New Ideas in Psychology, 28*(1), 79–93.
18. Guilford, J. P. (1967). *The nature of human intelligence*. New York: McGraw-Hill.
19. Hadamard, J. (1945). *The psychology of invention in the mathematical field.* New York: Dover Publications.

20. Haylock, D. W. (1987). A framework for assessing mathematical creativity in schoolchildren. *Educational Studies in Mathematics, 18*(1), 59–74.
21. Henry, J. (2009). Enhancing creativity with M.U.S.I.C. *The Alberta Journal of Educational Research, 5*(2), 199–211.
22. Huettel, S. A., Song, A. W., & McCarthy, G. (2009). *Functional magnetic resonance imaging* (2ed.). Massachusetts: Sinauer.
23. James, V., Lederman, G. R., & Vagt-Traore, B. (2004). Enhancing creativity in the classroom. In: M. Orey (Ed.), *Emerging perspectives on learning, teaching, and technology*. Athens, GA: Department of educational psychology and instructional technology, University of Georgia. Retrieved from http://www.coe.uga.edu/epltt/creativity.htm.
24. Jowett, B. M. A. (1937). *The dialogues of Plato*. New York: Random House.
25. Kandel, E. R. (2006). *In search of memory: The emergence of a new science of mind.* New York: W. W. Norton & Company.
26. Kaufman, J. C., & Beghetto, R. A. (2009). Beyond big and little: The four C model of creativity. *Review of General Psychology, 13*(1), 1–12.
27. Koichu, B., & Andžāns, A. (2009). Mathematical creativity and giftedness in out-of-school activities. In: R. Leikin, A. Berman & B. Koichu (Eds.), *Creativity in mathematics and education of gifted students* (285–308). Rotterdam, the Netherlands: Sense Publishers.
28. Kuhn, T. S. (1962). *Structure of scientific revolutions*. Chicago: The University of Chicago Press.
29. Leikin, R. (2009). Exploring mathematical creativity using multiple solution tasks. In: R. Leikin, A. Berman & B. Koichu (Eds.), *Creativity in mathematics and the education of gifted students* (129–145). Rotterdam, the Netherlands: Sense Publishers.
30. Liljedahl, P. (2008, February). *Mathematical creativity: In the words of the creators*. Proceedings of the 5th international conference on creativity in mathematics and the education of gifted students. Haifa, Israel.
31. Liljedahl, P., & Sriraman, B. (2006). Musings on mathematical creativity. *For The Learning of Mathematics, 26*(1), 20–23.
32. Lin, Y. S. (2011). Fostering creativity through education: A conceptual framework of creative pedagogy. *Creative Education, 2*(3), 149–155.

33. Lucas, B. (2001). Creative teaching, teaching creativity and creative learning. In: A. Craft, B. Jeffrey, & M. Leibling (Eds.), *Creativity in education* (35–44). London: Continuum.
34. Mann, E. L. (2006). Creativity: The essence of mathematics. *Journal for the Education of the Gifted, 30*(2), 236–260.
35. Martindale, C. (1989). Personality, situation, and creativity. In: J. A. Glover, R. R. Ronning, & C. R. Reynolds (Eds.), *Handbook of creativity: Perspectives on individual differences* (211–232). New York: Plenum Press.
36. McCrae, R. R. (1987). Creativity, divergent thinking, and openness to experience. *Journal of Personality and Social Psychology, 52*(6), 1258–1265.
37. Milgram, R. M., & Livne, N. L. (2005). Creativity as a general and a domain-specific ability: The domain of mathematics as an exemplar. In: J. C. Kaufman & J. Baer (Eds.), *Creativity across domains: Faces of the muse* (187–204). Mahwah, NJ: Lawrence Erlbaum and Associates.
38. Neumann, C. J. (2007). Fostering creativity — A model for developing a culture of collective creativity in science. *EMBO Reports, 8,* 202–206.
39. Piirto, J. (1999). *Talented children and adults, their development and education.* Upper Saddle River, NJ: Prentice Hall.
40. Pink, D. H. (2005). *A Whole new mind — Moving from the information age to the conceptual age*. New York: Penguin.
41. Plucker, J. A., Beghetto, R. A., & Dow, G. T. (2004). Why isn't creativity more important to educational psychologists? Potential, pitfalls, and future directions in creativity research. *Educational Psychologists, 39,* 83–96.
42. Polya, G. (1954). *Mathematics and plausible reasoning*. Princeton, NJ: Princeton University Press.
43. Sawyer, R. K. (1995). Creativity as mediated action: A comparison of improvisational performance and product creativity. *Mind, Culture, and Activity, 2,* 172–191.
44. Schindler, M., & Lilienthal, A. J. (2017). Eye-tracking as a tool for investigating mathematical creativity. Paper presented at the *10th International Conference on Mathematical Creativity and Giftedness* (MCG), Nicosia, Cyprus: Department of Education, University of Cyprus, 45–50

45. Schindler, M., Lilienthal, A. J., Chadalavada, R. T., & Ögren, M. (2016). Creativity in the eye of the student. Refining investigations of mathematical creativity using eye-tracking goggles. In: C. Csikos, A. Rausch, & J. Szitanyi (Eds.). *Proceedings of the 40th conference of PME — the international group for the psychology of mathematics education*, 4, 163–170. Szeged, Hungary.

46. Shriki, A. (2010). Working like real mathematicians: Developing prospective teachers' awareness of mathematical creativity through generating new concepts. *Educational Studies in Mathematics, 73,* 159–179.

47. Shriki, A., & Bar-On, E. (1997). Theory of global and local coherence and applications to geometry. In: E. Pehkonen (Ed.), *Proceedings of the 21st conference of the international group for the psychology of mathematics education* (Vol. 4, 152–159). Lahti, Finland.

48. Shriki, A., & Lavy, I. (2012). Teachers' perceptions of mathematical creativity and its nurture. In: T. Y. Tso (Ed.), *Proceedings of the 36th conference of the international group for the psychology of mathematics education* (Vol. 4, 91–98). Taipei, Taiwan: PME.

49. Shye, S., & Yuhas, I. (2004). *Creativity in problem solving: A multidimensional approach to its definition and measurement.* Jerusalem: The Van Leer Jerusalem Institute.

50. Silver, E. A. (1997). Fostering creativity through instruction rich in mathematical problem solving and problem posing. *ZDM — The International Journal on Mathematics Education, 29*(3), 75–80.

51. Simonton, D. K. (1999). *Origins of genius: Darwinian perspectives on creativity.* New York: Oxford University Press.

52. Smith, G. J. W. (2005). How should creativity be defined? *Creativity Research Journal, 17*(2–3), 293–295.

53. Sriraman, B. (2005). Are giftedness and creativity synonyms in mathematics? *The Journal of Secondary Gifted Education, XVII*(1), 20–36.

54. Sriraman, B. (2009). The characteristics of mathematical creativity. *ZDM — The International Journal on Mathematics Education, 41,* 13–27.

55. Sternberg, R. J. (1985). *Beyond IQ: A triarchic theory of intelligence.* Cambridge: Cambridge University Press.

56. Sternberg, R. J. (1997). A triarchic view of giftedness: Theory and practice. In: N. Colangelo & G. A. Davis (Eds.), *Handbook of gifted education* (43–53). Boston: Allyn and Bacon.
57. Sternberg, R. J., & Davidson, J. E. (2005). *Conceptions of giftedness.* Boston, MA: Cambridge University Press.
58. Sternberg, R. J., & Lubart, T. I. (1995). *Defying the crowd: Cultivating creativity in a culture of conformity*. New York: Free Press.
59. Sternberg, R. J., & Lubart, T. I. (1996). Investing in creativity. *American Psychologist, 51,* 677–688.
60. Sternberg, R. J., & Lubart, T. I. (2000). The concept of creativity: Prospects and paradigms. In: R. J. Sternberg (Ed.), *Handbook of creativity* (93–115). New York: Cambridge University Press.
61. Tatarkiewicz, W. (1980). *A history of six ideas: An essay in aesthetics*. Trans.: Christopher Kasparek. The Hague: Martinus Nijhoff.
62. Torrance, E. P. (1974). *Torrance tests of creative thinking*. Bensenville, IL: Scholastic Testing Service.
63. Torrance, E. P., Ball, E. O., & Safter, H. T. (1992). *Torrance Tests of Creative Thinking: Streamlined scoring guide Figural A and B*. Bensenville, IL: Scholastic Testing Service.
64. Wakefield, J. F. (1991). The outlook for creativity tests. *Journal of creative behavior, 25*(3), 184–193.
65. Wallas, G. (1926). *The art of thought*. New York: Harcourt, Brace.
66. Weisberg, R. W. (1993). *Creativity: Beyond the myth of genius*. New York: Freeman.
67. Welling, H. (2007). Four mental operations in creative cognition: The importance of abstraction. *Creativity Research Journal, 19*(2–3), 163–177.

https://doi.org/10.1142/9789811218750_0002

Chapter 2

Nurturing Mathematical Creativity at School

Atara Shriki

Introduction

The discussion of whether creativity can be nurtured through appropriate education has begun already in the 19th century when researchers have initiated an investigation of the relation between genius and creative achievements (Baer & Kaufman, 2005). For many years, researchers have been inclined to believe that creativity is innate and that an intended training has a limited impact, if any, on its development (Sternberg & Lubart, 1991). In the middle of the 20th century, inspired by studies in the field of psychometrics that were prevalent during the 1950s and 1960s, the perception of the source of creativity had undergone a gradual transformation, from being attributed only to geniuses to its representation through varied human skills. Today, many researchers concur that creativity can be nurtured by means of a suitable deliberate instruction (Fryer, 1996; Henry, 2009). Moreover, since creativity plays a major role as a motive for economic development (Lin, 2011) and for success in life in general (NACCCE, 1999), the changed perception evoked an interest in the development of creativity in educational frameworks (Esquivel, 1995).

Nevertheless, to date, there is still no consensus as to the way it can be done (Henry, 2009).

The first 'wave' of studies of creativity in education were conducted during the 1950s and 1960s. These studies engaged in pupil-oriented learning, while calling for a change in the traditional approach to teaching (Esquivel, 1995). Educators have believed that children are creative and curious by nature, open to experiences and attracted to new things. However, these features might gradually vanish if teachers do not generate suitable learning environments for them (Feldman & Benjamin, 2006; Torrance & Myers, 1970). The second 'wave' of studies of this field started in the 1990s, following extreme social, economic and technological fluctuations (Craft, 2001a). These studies focused on the need to nurture creativity in educational frameworks, maintaining that creativity facilitates survival and success in adult life (NACCCE, 1999). Based on this perception, the relation between creativity and education seems more important and essential than ever before, as this no longer concerns encouraging personal development and self-fulfillment. Rather, it relates to equipping the younger generation with instruments that will assist them later on in life. Attention should be paid to the fact that, regardless of the objective of nurturing creativity, studies in both 'waves' acknowledge that it is possible to nurture the creativity of every person (Lin, 2011).

In this regard, three major trends can be identified:

- Changing the perception of creativity, from viewing it as a personal trait that cannot be changed, to seeing it as a trait that can be developed through suitable guidance (Subotnik, Pillmeier, & Jarvin, 2009);

- Understanding the importance of developing the creativity of every learner and every teacher (Bolden, Harries, & Newton, 2010; Fryer, 1996); and

- Acknowledging the key role of the education system in the context of nurturing learners' creativity, as part of the wider responsibility to inculcate disciplinary knowledge (Lin, 2011).

All these have led educators and researchers to demonstrate a growing interest in the inquiry of creativity and in the appropriate ways for developing pupils' creativity (Henry, 2009). This interest has put on the agenda three main issues (Lin, 2011): instruction – ways of developing pupils' creativity and of offering them an opportunity for developing associated skills; learning environment – the characteristics of learning environments that can potentially support the development of creativity; and teachers' beliefs – beliefs regarding creativity and developing pupils' creativity as well as the ways intended to facilitate the nurturing and developing of pupils' creativity. These issues are discussed later on in this chapter.

Although teachers are expected to nurture the creative potential of their pupils (Baghetto, 2006) and most of the teachers acknowledge the importance thereof, efforts are usually not exerted in this respect at school (Sriraman, 2005). Consequently, pupils have only a very limited opportunity for experiencing creative learning and creative thinking (Silver, 1997). This reality is a result of several circumstances, among them: teachers' insufficient knowledge of this issue and beliefs that prevent teachers from providing their pupils with learning environments that can support the development of the latter's creativity (Bolden et al., 2010). Moreover, teachers are subject to a variety of external pressures, i.e. the need to cover the mandatory curriculum and help their pupils to succeed in external exams (Baghetto, 2006). In many cases, the education system highlights the importance of success in exams, and frequently teachers' quality is judged solely on the basis of their pupils' success. Generally, most of the mathematics external exams require a routine application of laws and algorithmic thinking (Sunderman & Kim, 2006) and the standard exams are not designed to assess creativity (Chamberlin & Moon, 2005). Hence, teachers tend to focus on memorization and algorithms, instead of nurturing their pupils' creativity.

Creativity and Instruction

The discussion of the relationship between instruction and creativity development has entailed, among others, a distinction between 'teaching creatively', 'teaching for creativity' and 'creative learning' (NACCCE, 1999). While 'teaching creatively' means using imaginative approaches to make learning more interesting, exciting and effective as well as developing materials and approaches that stimulate pupils' interests and motivate their learning, 'teaching for creativity' refers to forms of teaching that are intended to develop pupils' own creative thinking or behavior. Accordingly, instead of perceiving teaching and learning as independent processes that meet only seldom, these three elements should complement and strengthen each other. Thus, the interaction between the three of them will create a class climate that supports the development of creativity (Lin, 2011).

'Teaching creatively' requires that teachers use the power of their imagination in order to produce interesting and effective instructional approaches. On the other hand, 'teaching for creativity' relates to the objective of identifying pupils' creativity and stimulating it by offering opportunities for its nurture. These two approaches have a different starting point – teaching creatively focuses on teachers' work whereas teaching for creativity focuses on learners. However, in fact, these approaches cannot be separated. This is due to the fact that the features of teaching creatively, i.e. dynamic, innovative and full of imagination, stimulate learners' imagination, encouraging them to raise new ideas, lead directly to teaching for creativity (Jeffrey & Craft, 2004). Through creative instruction, teachers stimulate their pupils' creativity, catch up with their enthusiasm and inspire them to use their own imagination as well (Lucas, 2001). Such a way of instruction should provide pupils with adjusted time and space, encourage them to ask questions and turn them into active learners by transferring to them the responsibility for their own learning.

Learning environments designed for developing pupils' creativity

Based on the perception that creativity can be nurtured in educational frameworks, and in view of the relation between creativity, critical reflection and self-expression ability (Chamorro-Premuzic, 2006), pupils' creativity should be developed by generating learning environments that encourage creative thinking and support it (Eckhoff & Urbach, 2008).

The research literature engaging in learning environments designed for nurturing pupils' creativity, relates mainly to teachers' place in such environments and the characteristics of the activities to be included in them.

Torrance (1963) maintains that, in order to facilitate creative learning, pupils should be free of teachers' authority. Pupils learning under an authority, whereby teachers tell them what they must learn, what is the meaning of the ideas that they learn and how to learn them, cannot develop creative thinking. Conversely, eliminating the authority enables pupils to ask questions, investigate, look for answers, experience, and play. Children investigate out of inquisitiveness, a trait embodied in people since birth. Hence, pupils should be offered opportunities for experiencing it, while stop viewing teachers' role as 'knowledge delivery'. However, studies (e.g. Lin, 2011) indicate that many teachers find it hard to renounce their authority. Consequently, teachers' awareness of the need to give up part of their authority must be developed.

The development of pupils' creativity has to be supported by involving them in learning environments whereby they will engage in inquiry and discovery, without being restricted to specific rules or regulations (Mann, 2006). Pupils should be inspired to take an intellectual risk (Sriraman, 2009), and encouraged to cope with problems the solution of which is not immediate, thus necessitates to apply processes that are not routine or algorithmic (Haylock, 1987). In such cases, pupils are given the opportunity to associate between parts of their knowledge and create a whole and coherent knowledge, while applying processes that at first sight seem unrelated to the situation (NCTM, 2000).

Such processes are productive and assist in developing pupils' creativity, only if teachers manage to develop a learning environment that supports the pupils, promotes their confidence and respects both the pupils and the processes through which they conduct their independent inquiry. In such an environment, errors and incorrect answers are accepted openly, as part of the learning process, and are perceived as the outcome of the pupils' courage to create something (Goldin, 2008). Moreover, the learning environment should offer the pupils opportunities for sharing their knowledge. It is particularly important that teachers encourage pupils with creative thinking to share their ideas and insights (Sriraman, 2005). Examination of creative people's works illustrates that a large part of their breakthroughs has transpired as a result of collaboration with others and of social support (John-Steiner, 2000). Similarly, studies that monitored the conduct of creative organizations, have shown that a stimulus stemming from an interaction with colleagues is highly important (Neumann, 2007). Such an interaction supports the development of creativity since frequently an inspiration is evoked following the exchange of ideas. Sriraman (2009) investigated the nature of mathematical creativity by interviewing five leading mathematicians. Indeed, his study and the study of Liljedahl (2008) described in Chapter 1, indicate that all the mathematicians interviewed acknowledge the key role of social interaction in terms of the stimulus it provides to creative work. Moreover, a mathematical discourse about mathematics and mathematical ideas, the sharing of thoughts and presentation of arguments, enables pupils to critically reflect on their learning as well as reinforce their mathematics knowledge (NCTM, 2000).

Teachers should indeed challenge the pupils with tasks that require mathematical creativity and assess their creativity. However, teachers must also develop pupils' ability to assess their creativity on their own. Teachers have to discuss with their pupils about issues related to the assessment of thinking processes and outcomes. Such self-assessment in itself requires creativity as well as enables pupils to improve their solutions through an iterative process (Chamberlin & Moon, 2005).

Teachers' perceptions of creativity and development of pupils' creativity

Scheffler (1965) distinguished between teachers' beliefs and teachers' knowledge. Whereas beliefs are manifested by the tendency to act in a certain way in certain circumstances, knowledge allows actualizing the beliefs in practice. Consequently, teachers' perceptions are a combination of beliefs and knowledge, the latter consisting of disciplinary knowledge, pedagogical knowledge and class management knowledge (Shulman, 1987). According to Scheffler (1965), people, by their very nature, need no proof as to the "correctness" of their beliefs; beliefs, however, indicate a certain type of knowledge. Several researchers have engaged in the relation between teachers' perceptions of teaching and learning and their actions in practice (e.g. Kagan, 1992; Shulman, 1987; Thompson, 1992). They assumed that teachers' perceptions and actions are interrelated: the perceptions affect the quality of instruction and its nature, and the teaching process itself shapes the perceptions. Nevertheless, studies have found differences between teachers' perceptions or beliefs and their actual practice (e.g. Lev-Zamir & Leikin, 2011, 2013; Mohamed, 2006; Thompson, 1992). One of the reasons for the differences between beliefs about instruction and the instruction itself is related to the existence of two types of knowledge that affect the way teachers interpret their instruction. One type relates to issues of content knowledge, that is — curricular knowledge and knowledge of the way of presenting the lesson contents effectively and coherently. The second type is associated with teachers' instruction philosophy and understanding of the meaning of good instruction (Richards, 1996). Not always does this personal perspective, which encompasses beliefs and knowledge, has internal consistency. Alternately, Kagan (1992) maintains that the gap between the stated perceptions and the teaching in practice attests that a distinction should be made between 'superficial beliefs' (beliefs that are not truly part of teachers' educational philosophy) and 'deep beliefs' (beliefs that express a philosophy of education that is inherent in the teachers).

Lev-Zamir and Leikin (2013) distinguished between mathematics teachers' declarative perceptions, related mainly to the beliefs that they express, and actual perceptions that are reflected by their actions in class. In the context of mathematical creativity, the researchers made a distinction between two types of perceptions. The first is declarative perceptions relating to mathematics and referring to the way the teachers perceive the meaning of creative mathematical content. The second is perceptions of creative pedagogy, related to teachers' awareness of didactic and psychological aspects associated with creative learning. The study conducted by Lev-Zamir and Leikin (2013) illustrates that even in cases where teachers have the same declarative perceptions of mathematical creativity and ways of nurturing pupils' mathematical creativity, in practice they may display different behaviors. The gaps in behavior might stem from the way teachers translate their perceptions into action in practice, in particular, due to a difference between translating the perceptions into a teacher-oriented instruction and pupil-oriented instruction. While the first emphasizes flexibility in teaching but prevents flexibility in learning, the latter underscores flexibility and originality in creating solutions to a problem, with a clear intention to develop learners' creativity. Hence, Lev-Zamir and Leikin (2013) argue, we should examine teachers' declarative perceptions as to the development of pupils' creativity according to whether they are teacher-oriented or learner-oriented. The greater the extent to which the statements express more learner-oriented perceptions, the more we can expect from instruction, in general, to help pupils in developing their mathematical creativity. One of the assumptions presented by these researchers is that teachers' mathematics background plays a major role in translating the declarative perceptions of mathematical creativity into learner-oriented instruction. The researchers stipulate that when teachers have insufficient mathematics knowledge, they might not enable pupils to achieve flexibility in creating varied solutions for a problem.

Shriki and Lavy (2012) explored the perceptions and beliefs of 22 experienced middle and high school mathematics teachers about mathematical creativity, its nurture and assessment. Moreover, they examined the teachers' self-evidence regarding their relevant actual practice. The study indicates a gap between teachers' beliefs and their evidence with regard to the way they implement their pedagogical beliefs associated with creativity and its nurture. All the participant teachers maintained that it is important to develop pupils' mathematical creativity and that it is possible and necessary to foster mathematical creativity within the school framework. The teachers also argued that developing creativity facilitates mathematics learning in general, and contributes to changing the image of mathematics in the eyes of pupils. Although the teachers pointed out that they believed it was possible and necessary to nurture mathematical creativity at school, still half of them believed that not all the pupils are capable of displaying creative behavior. This position stands in stark contrast to the findings of previous findings (e.g. Isaksen, Dorval & Treffinger, 2000). The prevalent perception of the teachers who participated in the study was creativity from the aspect of the product and, in particular, non-routine solution/explanation/proof. That is, like the common image of creativity (Leikin, 2009; Shriki, 2010), most of the participants perceived the meaning of a creative product in the context of its originality. They perceived an 'original product' as "a proof that I have never seen", "a solution that surprised me", or "an explanation that has never crossed my mind". Attention should be paid to the fact that this perception of the meaning of the original product is problematic since it implies that teachers judge the originality level of the pupils' products while relating to themselves as the reference group. This finding can account for teachers' tendency to see a limited group of pupils as capable of producing a creative outcome. Hence, if teachers are expected to believe that all pupils can demonstrate creativity, they should understand the meaning of personal creativity and examine the process of personal creativity development of each and every pupil. Responding to the question that engaged in the steps adopted by teachers in practice in order to encourage the development of their pupils' creativity, about 25% of them admitted they were making no intentional action for that purpose. The rest answered that they were developing mathematical creativity by occasionally presenting to the pupils' examples of solving a problem in more than one way. This was done mainly in order to expose the pupils to the relations between the various fields of mathematics. Nevertheless, even teachers who from time to time expose their pupils to different solutions, admitted that they do not explicitly ask the pupils to do it as well. Moreover, they do not ask them to discuss the differences between the various ways of solving the problem.

The teachers attributed the main reason for not encouraging pupils to solve problems in various ways as a matter of routine mainly to lack of time and their pressure to prepare the pupils for external exams. They argued that in such exams, the pupils are anyhow not required to present non-routine solutions, hence education for creativity is a 'luxury'. Furthermore, some teachers maintain that if they ask the pupils to solve a problem in various ways, this will take too much time of the lesson since they will have to allow the pupils to present their various solutions. Apparently, teachers perceive themselves mostly as 'delivering material'. Therefore, although they acknowledge that it is important to encourage pupils to solve problems in various ways and recognize its relation to the development of pupils' mathematical creativity, they tend to avoid it due to lack of time and inability to view the development of the pupils' mathematical creativity as part of the standard teaching and learning process.

The teachers were also asked to relate to the issue of namely rewarding pupils who produced creative outcomes, for instance, a non-routine solution/explanation/proof. Some of the teachers argued that creative solutions on the test should be encouraged by accreditation of bonus points. Others believed that pupils who for should be praised in class, stimulating them (and possibly others) to continue in that direction. Conversely, some of the teachers believed that since only a few pupils are capable of demonstrating mathematical creativity, it is unjust to discriminate pupils on these grounds. These teachers believed that the comparison between pupils should be done only on the basis of differences in knowledge, as measured by the exam scores. According to them, teachers are not meant to examine whether pupils are creative or not, but rather examine their level of understanding of the learning material. To sum up, Shriki and Lavy (2012) concluded that teachers' insufficient knowledge regarding creativity and the ways of nurturing and assessing it, is the main reason for their refrain from generating learning environments that support the development of pupils' mathematical creativity.

Several international comparative studies have been conducted in the context of teachers' beliefs about mathematical creativity. The literature review of Bolden et al. (2010) illustrates that teachers in various locations around the globe, perceive creativity in quite a similar way. A variety of local and comparative studies in this regard have been conducted with teachers from England, United States, India, Cyprus, Sweden, and other countries. Common to all these studies was the finding that teachers perceive creativity as a process of independent work, leading to a creative outcome. Teachers tend to define creative pupils as ones with rich imagination. Yet, they associated creativity mainly with the production of an outcome in the field of art and music. In other words, the prevalent perception of mathematics teachers is that teaching mathematics offers only a few opportunities for developing pupils' creativity, particularly in older age groups. In addition, they sometimes perceive the focus on creativity as a distraction from the learning.

Leikin, Subotnik, Pitta-Pantazi, Singer, and Pelczer (2013) conducted a study of cultural aspects of creativity. To that end, they asked about 1100 high school mathematics teachers from six countries (India, Israel, Latvia, Mexico, Cyprus, and Romania) to respond to a questionnaire. The questionnaire comprised 100 assertions and related to teachers' perceptions of aspects such as: who is a creative person and who is a mathematically creative pupil, who is a mathematically creative teacher, and how is mathematical creativity connected to culture. Generally speaking, the findings show that, in spite of the cultural and traditional differences between the countries that have participated in the study, the variation between the teachers by countries is low. Leikin et al. (2013) concluded that creativity in mathematics education is probably an inter-cultural issue. The broadest agreement among teachers from the different countries related to the unique characteristic of creative pupils (e.g.

abilities to conduct an inquiry, mathematical flexibility, and successful problem solution). They also concurred about the characteristics of mathematically creative teachers (e.g. pleasure of teaching mathematics, ability to inspire pupils' creativity and a willingness to assess it, and linking mathematical contents to other areas of knowledge). In particular, the teachers agreed that teaching creatively means teaching mathematics in a flexible manner according to a pupil-oriented approach as well as planning in a flexible way the mathematics lessons themselves. The differences between the teachers' perceptions were manifested by several of the indices which, according to the researchers, could be attributed to differences associated with the education system and culture (Leikin et al., 2013). Thus, for example, much more than the other teachers, the Romanian teachers agreed with most of the assertions that described characteristics of a creative person in general and of mathematically creative pupils in particular. Moreover, they agreed with the assertions that related to the characteristics of creative teachers. Leikin et al. (2013) stipulated that this is due to the over-emphasis put in Romania on mathematical creativity and to the fact that mathematics teacher education is oriented at excellence in mathematics. Teachers from India and Mexico agreed more than the other teachers with assertions engaging in the creativity-culture relation. Apparently, this stems from the historical-cultural characteristics of these countries in comparison with the other four participating countries.

Creativity in Teaching Mathematics at School

Mathematical creativity guarantees, in fact, the development of the entire field of mathematics. The continuous increase in the number of journals that publish studies of mathematics attests that this field is constantly developing. Nevertheless, the driving force for this development, namely mathematicians' creativity, has not been sufficiently explored (Sriraman, 2009).

Although it is reasonable to assume that researchers in the field of mathematics education would pay attention to issues related to the nurture of pupils' mathematical creativity, studies of this field hardly engage in creativity (Lev-Zamir & Leikin, 2011). This is attested by the empirical literature review performed by Haylock (1987) regarding the relation between creativity and mathematics education. This review indicates that between the years 1966–1985, there was only a limited reference to this issue in the empirical literature. Leikin (2009, 2011) performed a similar review, showing that between the years 1999–2009, this field was not sufficiently investigated as well.

The empirical literature engaging in mathematical creativity is divided today into several foci of interest. Among them are perceptions of pre-service and in-service teachers related to the meaning of mathematical creativity, manifestations of creative ability in mathematics and ways of nurturing pupils' mathematical creativity, as well as diagnosis of pupils' mathematical creativity in the context of problem solving and problem posing. It is rather surprising that there are no comprehensive studies that investigate the characteristics of learning environments that have the potential to develop pupils' mathematical creativity.

Teachers' beliefs about the meaning of teaching creatively and teaching for creativity have implications for the way they are implementing it in practice (Bolden et al., 2010). Thus, it is highly important to explore the beliefs and perceptions of pre-service and in-service teachers with regard to these issues. First, if teachers are expected to stimulate the development of mathematical creativity and support such a process, they must first and foremost acknowledge that creativity is not a given or fixed trait of exceptionally gifted pupils. It is rather a dynamic trait that can be developed among all pupils through appropriate instruction (Beghetto & Kaufman, 2009; Leikin, 2007; Liljedahl & Sriraman, 2006; Shriki, 2010). Such instruction requires, among others, that teachers emphasize the deepening of mathematics knowledge so that it supports pupils' ability to draw conclusions flexibly and assists in the developing of pupils' divergent thinking (Silver, 1997). Divergent thinking can be developed by means of instruction that underscores the fluency and flexibility in problem solving, as opposed to traditional instruction that mainly encourages the development of convergent thinking, oriented at attaining a single solution or mathematical idea (Kwon, Park, & Park, 2006; Silver, 1997).

Although most mathematics teachers acknowledge the importance of developing their pupils' creativity (e.g. Shriki, 2010; Shriki & Lavy, 2012), the empirical literature illustrates that usually, pupils do not get to study in environments that nurture their mathematical creativity

(Sriraman, 2005). Consequently, pupils have few opportunities for experiencing an instruction that encourages creative thinking (Silver, 1997). Unlike professional mathematicians, who mainly engage in problems whereby more is unknown than known, at school, both the curriculum and the pedagogical approaches hardly facilitate the pupils' encounter with the open aspect of mathematics. This is due to the fact that most teachers avoid presenting open or incomplete problems for the pupils. Thus, they prevent the pupils from engaging over time in problems of this kind. Frequently, when pupils are requested to apply a certain approach systematically in order to solve a problem, they do not need at all to implement creative thinking (Haylock, 1987). In the terms of Guilford (1967), pupils are required to apply in fact only convergent thinking. Avoiding the development of mathematical creativity at school, pupils get the impression that mathematics is nothing but a collection of skills and rules to be memorized. Hence, their natural inquisitiveness about mathematics and their enthusiasm for it might fade out (Mann, 2006; Shriki, 2004).

One of the main reasons for teachers' avoidance of generating learning environments designed to nurture pupils' creativity, resides in the prevalent educational reality. In this reality, teachers tend to teach in a way similar to the one they have been taught during their school years (Hall, Fisher, Musanti, & Halquist, 2006). As pupils, most of them experienced mathematics education that did not highlight the development of creativity. Consequently, in the absence of a suitable role model, it is likely to assume that as teachers they do not have sufficient tools or appropriate methods for nurturing their pupils' creativity. Another problem is the lack of suitable learning materials, rendering it difficult for teachers to initiate learning environments designed for the nurture of creativity (Silver, 1997). Moreover, in the absence of a simple and available instrument for assessing creativity, teachers find it hard to assess their pupils' mathematical creativity and monitor the development thereof (Shriki, 2010). Many teachers are also apprehensive that nurturing pupils' creativity will take too much time at the expense of teaching the binding mathematics curriculum (Levav-Waynberg & Leikin, 2012). This is due to the fact that they are not aware that mathematical creativity can be nurtured simultaneously with the instruction of the required contents.

Accordingly, in what follows we discuss two of the possible ways of nurturing mathematical creativity as an integral part of the curriculum. These are solving problems in various ways and problem posing. While contemporary perceptions of creativity differ one from the other in relation to the essence of creativity and the ability of individuals to produce a creative outcome, there is an increasing consensus related to the centrality of problem solving and problem posing processes within the creative mathematical practice (Silver, 1997). These ways do not oblige teachers to use unique instruction materials and therefore, they can generate learning environments that encompass these components, according to the pupils' level of knowledge and to relevant learning topics.

Nurturing mathematical creativity by encouraging pupils to solve problems in various ways

Many researchers link mathematical creativity and the ability to solve a problem in various ways (e.g. Ervynck, 1991, Kwon et al., 2006; Leikin, 2009; Silver, 1997), as this ability attests to divergent thinking.

Schoenfeld (1989) defined the meaning of a 'mathematical problem' as a task that obliges pupils to make some sort of decision, although they have no available direct means that allows them to do it. A mathematical task to which pupils have an available means allowing them to make a decision about it in a direct manner is perceived as an 'exercise' and not as a 'problem'. The professional literature frequently refers to these problems as 'insight problems' (Sternberg & Davidson, 1995). Problems of this kind have a direct relation to creativity since they require the learners to undergo a process of incubation and insight due to lack of direct or algorithmic solutions.

Teachers' main tendency is to assess their pupils' mathematical creativity by referring to a product, mainly one that is an original solution of a given problem (Leikin, 2009; Shriki, 2010; Shriki & Lavy, 2012), a new approach to the examination of a familiar problem (Liljedahl & Sriraman, 2006), or solution of a non-routine problem (Chiu, 2009). We will, therefore, discuss below mainly aspects associated with the creative product and its assessment rather than the creative process itself.

In order to assess the creativity of pupils coping with the solution of problems in various ways, one can apply the indices proposed by Torrance

(1974). He related to four dimensions for the assessment of creativity: fluency, flexibility, novelty, and elaboration. Researchers of mathematical creativity adjusted these dimensions to the context of problem solving in various ways. Drawing on Torrance (1974), Silver (1997) defined 'fluency' as the generation of multiple ideas or multiple solutions to a given problem, 'flexibility' as finding solutions based on different ideas and 'novelty' as the investigation of various solutions to a problem and generation of a new solution grounded in these solutions.

On this basis, Leikin (2009) developed an instrument allowing the examination of mathematical creativity in the context of multiple solutions to a given problem. She suggests assessing mathematical creativity by means of tasks that require multiple solutions, where "*Solutions to the same problem are considered to be different if they are based on: (a) different representations of some mathematical concepts involved in the task, (b) different properties (definitions or theorems) of mathematical objects within a particular field, or (c) different properties of a mathematical object in different fields*" (p. 133). Furthermore, this instrument relates to the fact that creativity is relative and not absolute and, hence, pupils' creativity must be assessed according to their reference group. In this context, Leikin (2009) coined the terms 'solution spaces', 'individual solution spaces' and 'collective solution spaces'. As far as a given problem is concerned, 'solutions space' is the collection of solutions produced by experts in the field and it comprises all the known solutions of a certain problem at a particular time. On the other hand, the pupils' routine solutions space consists of the solutions familiar to them from the curriculum, from the textbooks and from the teachers' presentations. The "individual solutions space" constitutes the collection of solutions that pupils create in response to a given problem. This space can be characterized by a space referred to by Leikin (2009) as the 'available individual solution space', encompassing solutions that pupils can attain by themselves as well as by the 'potential solution space', comprising solutions that pupils can attain by external assistance. The 'collective solution space' is a combination of all the solutions produced by a certain pupil group. Such spaces are usually embodying more solutions than the 'individual solution spaces' of each of the group members, constituting a source for the development of their individual solution spaces. Reference to the individual solution space facilitates the assessment of each and every pupil's creativity according to the fluency and flexibility dimensions. Comparison of the individual solution space and the collective solution space allows assessment of the solutions' extent of originality. The instrument proposed by Leikin (2009) for the assessment of pupils' creativity can also be used by the teachers for developing pupils' creativity in the context of solving mathematical problems in various ways. Moreover, it can assist teachers in examining how they develop their pupils' creativity (Levav-Waynberg & Leikin, 2012).

Nurturing mathematical creativity by means of tasks that require solving problems in various ways, depends, to a large extent, on teachers' willingness. Levav-Waynberg & Leikin (2012) have found that mathematics teachers are more willing to implement such learning processes when geometry instruction is concerned. The researchers argue that geometry offers a variety of opportunities for inquiry activities. Moreover, many geometric problems naturally enable solutions in various ways, since most problems do not lead to algorithmic or procedural solutions, unlike other areas of school mathematics. According to Levav-Waynberg and Leikin (2012), teachers believe that there is a relation between solving geometric problems in various ways, particularly at high levels and developing mathematics knowledge. Furthermore, teachers do not consider it as a task that comes at the expense of the time allocated to the instruction of the curriculum.

Nurturing mathematical creativity by encouraging pupils to pose problems and questions

Posing problems is associated with the creation of new problems and inquiry questions about a given situation as well as the reformulation of a problem while solving it (Silver, 1994). Mathematicians do solve problems posed by others. However, they usually formulate the problems by themselves on the basis of personal experiences and interests. Conversely, as far as school mathematics is concerned, in most cases, the problems are presented by the teachers and the textbooks (Silver, 1997).

Albert Einstein believed that "*The formulation of a problem is often more essential than its solution, which may be merely a matter of mathematical or experimental skills. To raise new questions, new possibilities, to regard old questions from a new angle, requires creative imagination and marks a real advance in science*" (Einstien & Infeld, 1938, in Ellerton & Clarkson, 1996, p. 518). Similar to Einstein, Hadamard (1945) also referred to the ability to raise questions as an indicator of an exceptional mathematical talent. The history of mathematics does, in fact, attest that many papers contributed to the development of mathematics, not just because their writers managed to prove some theorem. Rather, it was due to the fact that they presented a problem that opened a window to a completely new research area in mathematics (Van Harpen & Sriraman, 2013).

Pupils are traditionally required to solve the problems that teachers present to them, those problems the teachers believe are relevant to the pupils' learning (English, 1997). On the other hand, they hardly engage in an activity of problem posing (Kojima & Miwa, 2008). The didactic ideas underpinning the reform in mathematics education illustrate the need for transferring part of the responsibility for the learning to the pupils themselves, enabling them, among others, to raise questions they deem worthy of discussion (English, 1997). The transition from solving problems suggested by the teachers to posing problems by the pupils is, in fact, a transition from 'telling the pupils what they have to do' to 'listening to what the pupils have to say' (NCTM, 2000).

In order to solve problems and pose problems, pupils have to use similar cognitive actions. Consequently, researchers maintain that posing problems has a positive effect on pupils' ability to solve problems (e.g. Nikita & Shimada, 2005; Silver & Cai, 1993). Nevertheless, there are differences between these two actions. On the one hand, problem posing relies on the process of problem solving. For example, identifying the main components of a problem and their interrelations as well as identifying the idea to which we have to aspire (e.g. proving a theorem). On the other, problem posing takes the pupils beyond the parameters of the solution process. For example, problem posing engages the pupils with questions about the source of the problem ideas or evokes thinking about other problems that might be posed if the components of the original problem are changed or expanded (English, 1997).

Developing pupils' ability to raise questions might reduce their dependence on the teachers and the textbooks, encourage them to assume responsibility for their learning and develop their skills as independent learners. When pupils formulate new problems, they develop a sense of ownership on the learning material, reinforcing the inquisitiveness and enthusiasm about the learning of mathematics (Cunningham, 2004). Posing problems develops an investigation ability and divergent and flexible thinking. Moreover, it makes teachers and pupils aware of misconceptions, creates an image of mathematics as a dynamic and growing discipline and reduces mathematics anxiety (Brown & Walter, 1969, 1990; English, 1996). When pupils pose problems, they are also reinforcing and enriching their mathematics knowledge, since, in order to be able to pose a valid problem, they should be aware of the implicit relations hidden in the situation of the given problem (Lavy & Shriki, 2010; Shriki & Lavy, 2012).

And what about the relation between problem posing and creativity? In 1993, Edward Silver gave a plenary lecture within the framework of the International Group for the Psychology of Mathematics Education (PME) that took place in Japan (Silver, 1994). In his lecture, he commented that we still do not know whether there is a clear relation between creativity and the ability to pose problems. For the past 25 years since that lecture, researchers have attempted to investigate the existence of this relation and its essence thereof. Some of them (e.g. Jay & Perkins, 1997) stipulate that in the context of developing creativity, problem posing is more important than problem solving. However, the empirical literature that engages in the relation between creativity and solving problems in various ways is richer than that dealing with the relation between creativity problem posing. Hence, the essence of the relationship between problem solving and creativity is not sufficiently clear (Van Harpen & Sriraman, 2013). Nevertheless, according to Koichu & Kontorovich, 2013), the empirical literature that discusses the posing of mathematical problems, indicates that the educational opportunities embodied in this activity are more extensive and powerful than the activities encompassed in problem solving. Thus, problem posing is considered as an effective research instrument for the exposure of mathematics knowledge components and as a means of developing creative mathematical thinking. Yet, many studies of this field have illustrated that in-service teachers, pre-service teachers, and pupils fail to pose problems with an exceptional meaning and, frequently, the problems are not even properly formulated. For the time being, therefore, information regarding processes involved in problem posing is insufficiently clear.

Similar to the case of solving a problem in various ways and creativity, researchers applied the indices proposed by Torrance (1974) also for the examination of the relationship between problem posing and creativity. As will be demonstrated in 10, fluency is measured according to the number of various problems that were posed; flexibility is measured according to the number of different problem categories; novelty is measured in accordance with the relative frequency of the problems in the reference group to which the pupils belong; and elaboration is measured according to the number of problems formulated as a generalization (Koichu, 2008; Shriki, 2013; Shriki & Lavy, 2010; Silver, 1997). To date, only a few studies of this field have been conducted.

References

1. Baer, J., & Kaufman, J. C. (2005). Bridging generality and specificity: The Amusement Park Theoretical (APT) Model of creativity. *Roeper Reviewer, 27,* 158–163.
2. Beghetto, R. A. (2006). Creative justice? The relationship between prospective teachers' prior schooling experiences and perceived importance of promoting pupil creativity. *The Journal of Creative Behavior, 40*(3), 149–162.
3. Beghetto, R. A., & Kaufman, J. C. (2009). Do we all have multicreative potential? *ZDM - The International Journal on Mathematics Education, 41,* 39–44.
4. Bolden, D. S., Harries, T. V., & Newton, D. P. (2010). Pre-service primary teachers' conceptions of creativity in mathematics. *Educational Studies in Mathematics, 73*(2), 143–157.
5. Brown, S. I., & Walter, M. I. (1969). What if not? *Mathematics Teaching, 46,* 38–45.
6. Brown, S. I., & Walter, M. I. (1990). *The art of problem posing* (2nd ed.). Hillsdale, NJ: Lawrence Erlbaum Associates.
7. Chamberlin, S. A., & Moon, S. M. (2005). Model-eliciting activities as tool to develop and identify creatively gifted mathematicians. *Journal of Secondary Gifted Education, 17*(1), 3–47.
8. Chamorro-Premuzic, T. (2006) Creativity versus conscientiousness: Which is a better predictor of pupil performance? *Applied Cognitive Psychology, 20,* 521–531.
9. Chiu, M. S. (2009). Approaches to the teaching of creative and non-creative mathematical problems. *International Journal of Science and Mathematics Education, 7,* 55–79.
10. Craft, A. (2001a). *An analysis of research and literature on creativity in education.* Report prepared for the Qualifications and Curriculum Authority. Retrieved from http://www.euvonal.hu/images/creativity_report.pdf

11. Cunningham, R. (2004). Problem posing: An opportunity for increasing pupil responsibility. *Mathematics and Computer Education, 38*(1), 83–89.
12. Eckhoff, A., & Urbach, J. (2008). Understanding imaginative thinking during childhood: Sociocultural conceptions of creativity and imaginative thought. *Early Childhood Education Journal, 36,* 179–185.
13. Ellerton, N. F., & Clarkson, P. C. (1996). Language factors in mathematics teaching and learning. In A. I. Bishop (Ed.), *International handbook of mathematics education* (987–1033). Netherlands: Kluwer Academic Publishers.
14. English, L. (1996). Children problem posing and problem-solving preferences. In: J. Mulligan & M. Nitchelmore (Eds.), *Research in early number learning* (227–242). Adelaide, Australia: Australian Association of Mathematics Teachers.
15. English, L. D. (1997). Promoting a problem-posing classroom. *Teaching Children Mathematics*, *4*(3), 172–179.
16. Ervynck, G. (1991). Mathematical creativity. In: D. Tall (Ed.), *Advanced mathematical thinking* (42–53). Dordrecht, Netherlands: Kluwer.
17. Esquivel, G. (1995). Teacher behaviors that foster creativity. *Educational Psychology Review, 7*(2), 185–202.
18. Feldman, D. H., & Benjamin, A. C. (2006). Creativity and education: An American retrospective. *Cambridge Journal of Education, 36,* 319–336.
19. Fryer, M. (1996). *Creative teaching and learning*. London: Paul Chap-man Publishing Ltd.
20. Goldin, G. A. (2008). The affective dimension of mathematical inventiveness. In: R. Leikin (Ed.), *Proceedings of the 5th international conference on creativity in mathematics and the education of gifted pupils* (1–14). Haifa, Israel: CET - The Center of Educational Technology.
21. Guilford, J. P. (1967). *The nature of human intelligence*. New York: McGraw-Hill.
22. Hadamard, J. (1945). *The psychology of invention in the mathematical field.* New York: Dover Publications.
23. Hall, L. D., Fisher, C., Musanti, S., & Halquist, D. (2006). Professional development in teacher education: What can we learn from PT3? *Tech Trends, 50*(3), 25–31.
24. Haylock, D. (1997). Recognizing mathematical creativity in school children. *International Reviews on Mathematical Education, 29*(3), 68–74. Retrieved from http://www.fiz-karlsruhe.de/fix/publications/zdm/adm97

25. Henry, J. (2009). Enhancing creativity with M.U.S.I.C. *The Alberta Journal of Educational Research, 5*(2), 199–211.
26. Isaksen, S., Dorval, K., & Treffinger, D. (2000). *Creative approaches to problem solving*. Dubuque, Iowa: Kendall/Hunt Publishing Company.
27. Jay, E. S., & Perkins, D. N. (1997). Problem finding: The search for mechanism. In: M. Runco (Ed.), *The creativity research handbook* (257–293). New Jersey: Hampton Press.
28. Jeffrey, B., & Craft, A. (2004). Teaching creatively and teaching for creativity: Distinctions and relationships. *Educational Studies, 30,* 77–87.
29. John-Steiner, V. (2000). *Creative collaboration.* New York: Oxford University Press.
30. Kagan, D. M. (1992). Implication of research on teacher belief. *Educational Psychologist, 27*(1), 65–90.
31. Koichu, B. (2008). If not, what yes? *International Journal of Mathematical Education in Science and Technology, 39*(4), 443–454.
32. Koichu, B., & Kontorovich, I. (2013). Dissecting success stories on mathematical problem posing: A case of the Billiard Task. *Educational Studies in Mathematics, 83*(1), 71–86.
33. Kojima, K., & Miwa, K. (2008). A System that facilitates diverse thinking in problem posing. *International Journal of Artificial Intelligence in Education, 18*(3), 209–236.
34. Kwon, O. N., Park, J. S., & Park, J. H. (2006). Cultivating divergent thinking in mathematics through an open-ended approach. *Asia Pacific Education Review, 7,* 51–61.
35. Lavy, I., & Shriki, A. (2010). Engaging in problem-posing activities in a dynamic geometry setting and the development of prospective teachers' mathematical knowledge. *Journal of Mathematical Behavior, 29,* 11–24.
36. Leikin, R. (2007). Habits of mind associated with advanced mathematical thinking and solution spaces of mathematical tasks. In: D. Pitta-Pantazi & G. Philippou (Eds.), *Proceedings of the fifth conference of the European Society for Research in Mathematics Education - CERME-5* (2330–2339). Retrieved from http://ermeweb.free.fr/Cerme5.pdf
37. Leikin, R. (2009). Exploring mathematical creativity using multiple solution tasks. In: R. Leikin, A. Berman & B. Koichu (Eds.), *Creativity in mathematics and the education of gifted pupils* (129–145). Rotterdam, the Netherlands: Sense Publishers.

38. Leikin, R. (2011). The education of mathematically gifted pupils: On some complexities and questions. *Montana Mathematical Enthusiast Journal*, *8*, 167–188.

39. Leikin, R., Subotnik, R., Pitta-Pantazi, D., Singer, M., & Pelczer, I. (2013). Teachers' views on creativity in mathematics education: An international survey. *ZDM - The International Journal on Mathematics Education, 45*(2), 309–324.

40. Lev-Zamir, H., & Leikin, R. (2011). Creative mathematics teaching in the eye of the beholder: Focusing on teachers' conceptions. *Research in Mathematics Education, 13*(1), 17–32.

41. Lev-Zamir, H., & Leikin, R. (2013). Saying versus doing: Teachers' conceptions of creativity in elementary mathematics teaching. *ZDM - The International Journal on Mathematics Education, 45*(2), 295–308.

42. Levav-Waynberg, A., & Leikin, R. (2012). Using multiple solution tasks for the evaluation of pupils' problem-solving performance in geometry. *Canadian Journal of Science, Mathematics and Technology Education, 12*(4), 313–333.

43. Liljedahl, P. (2008, February). Mathematical creativity: In the words of the creators. *Proceedings of the 5th international conference on creativity in mathematics and the education of gifted pupils*. Haifa, Israel.

44. Liljedahl, P., & Sriraman, B. (2006). Musings on mathematical creativity. *For The Learning of Mathematics, 26*(1), 20–23.

45. Lin, Y. S. (2011). Fostering creativity through education: A conceptual framework of creative pedagogy. *Creative Education, 2*(3), 149–155.

46. Lucas, B. (2001). Creative teaching, teaching creativity and creative learning. In: A. Craft, B. Jeffrey, & M. Leibling (Eds.), *Creativity in education* (35–44). London: Continuum.

47. Mann, E. L. (2006). Creativity: The essence of mathematics. *Journal for the Education of the Gifted, 30*(2), 236–260.

48. Mohamed, N. (2006). *An exploratory study of the interplay between teachers' beliefs, instructional practices & professional development*. Unpublished doctoral dissertation, the University of Auckland. Retrieved from http://www.asian-efl-journal. com/ Thesis_ Naashia.pdf

49. NACCCE - National Advisory Committee on Creative and Cultural Education (1999). *All our futures: Creativity, culture and education.* London: Department for Education and Employment. Retrieved from http://sirkenrobinson.com/pdf/allourfutures.pdf

50. NCTM - National Council of Teachers of Mathematics (2000). *Principles and standards for school mathematics*. Reston, VA: NCTM.

51. Neumann, C. J. (2007). Fostering creativity - A model for developing a culture of collective creativity in science. EMBO Reports, 8, 202–206.

52. Nikata, K., & Shimada, H. (2005). Facilitation of analogical transfer by posing an analogous problem for oneself. *Japanese Journal of Educational Psychology, 53,* 381–392.

53. Richards, J. C. (1996). Teachers' maxims in language teaching. *TESOL Quarterly, 30*(2), 281–296.

54. Scheffler, I. (1965). *Condition of knowledge: An introduction to epistemology and education.* Chicago: The University of Chicago Press.

55. Schoenfeld, A. H. (1989). Teaching mathematical thinking and problem solving. In: L. B. Resnick & L. E. Klofer (Eds.), *Towards the thinking curriculum: Current cognitive research* (83–103). Alexandria, VA: Association for Supervision and Curriculum Development.

56. Shriki, A. (2004). Everyone can be a real mathematician. *ALEH- Journal for Mathematics Teachers, 31*, 4–13. (In Hebrew)

57. Shriki, A. (2010). Working like real mathematicians: Developing prospective teachers' awareness of mathematical creativity through generating new concepts. *Educational Studies in Mathematics, 73,* 159–179.

58. Shriki, A. (2013). A model for assessing the development of pupils' creativity in the context of problem posing. *Creative Education, 4*(7), 430–439.

59. Shriki, A., & Lavy, I. (2010, August). *Evaluating the creativity of pupils' products in geometry.* Paper presented at the 6th international conference on creativity in mathematics and the education of gifted pupils, Riga, Latvia.

60. Shriki, A. & Lavy, I. (2012). Problem posing in a dynamic geometry environment and the development of mathematical insights. *The International Journal of Learning, 18*(5), 61–70.

61. Shulman, L. S. (1987). Knowledge and teaching: Foundations of the new reform. *Harvard Educational Review, 56,* 1–22.
62. Silver, E. A. (1994). On mathematical problem posing. *For the Learning of Mathematics, 14,* 19–28.
63. Silver, E. A. (1997). Fostering creativity through instruction rich in mathematical problem solving and problem posing. *ZDM - The International Journal on Mathematics Education, 29*(3), 75–80.
64. Silver, E. A, & Cai, J. (1993). Mathematical problem posing and problem solving by middle school pupils. In: C. A. Maher, G. A. Goldin & R. B. Davis (Eds.), *Proceedings of PMENA 11* (Vol. 1, 263–269). New Brunswick, NJ: Rutgers University.
65. Sriraman, B. (2005). Are giftedness and creativity synonyms in mathematics? *The Journal of Secondary Gifted Education, XVII*(1), 20–36.
66. Sriraman, B. (2008). Are mathematical giftedness and mathematical creativity synonyms? A theoretical analysis of constructs. In: B. Sriraman (Ed.), *Creativity, giftedness, and talent development in mathematics* (85–112). Charlotte, NC: Information Age Publishing, INC.
67. Sriraman, B. (2009). The characteristics of mathematical creativity. *ZDM - The International Journal on Mathematics Education, 41,* 13–27.
68. Sternberg, R. J., & Davidson, J. E. (Eds.). (1995). *The nature of insight.* Cambridge, MA: MIT Press.
69. Sternberg, R. J., & Lubart, T. I. (1991). An investment theory of creativity and its development. *Human Development, 34,* 1–31.
70. Subotnik, R. F., Pillmeier, E., & Jarvin, L. (2009). The psychosocial dimensions of creativity in mathematics: Implications for gifted education policy. In: R. Leikin, A. Berman & B. Koichu (Eds.), *Creativity in mathematics and the education of gifted pupils* (165–179). Rotterdam, the Netherlands: Sense Publishers.
71. Sunderman, G. L., & Kim, J. S. (2006). Implementing supplemental educational services: Implications for school districts and educational opportunity. In: K. K. Wong & S. A. Rutledge (Eds.), *System-wide efforts to improve pupil achievement* (72–83). Greenwich, CT: Information Age Publishing, Inc.

72. Thompson, A. (1992). Teachers' beliefs and conceptions: A synthesis of the research. In: D. Grouws (Ed.), *Handbook of research on mathematics teaching and learning* (127–146). New York: Macmillan.
73. Torrance, E. P. (1963). Creativity. In: F. W. Hubbard (Ed.), *What research says to the teacher* (Number 28). Washington, DC: Department of Classroom Teachers American Educational Research Association of the National Education Association.
74. Torrance, E. P., & Myers, R. E. (1970). *Creative learning and teaching*. New York: Harper & Row.
75. Van Harpen, X. Y., & Sriraman, B. (2013). Creativity and mathematical problem posing: An analysis of high school pupils' mathematical problem posing in China and the USA. *Educational Studies in Mathematics, 82*(2), 201–221.

Section II

Creative Approaches to the Teaching of Geometry

Photo by S. Gazit

https://doi.org/10.1142/9789811218750_0003

Chapter 3

Creative View of Interior and Exterior Angles of Convex Polygons

Ilana Levenberg & Dorit Patkin

*Geometry is a **skill** of the **eyes** and the **hands** as well as of the **mind** / Jean Pedersen,*

Introduction

When starting to study geometry, learners are presented with concepts, such as lines, angles, polygons, and diagonals. These concepts can be displayed in various ways, aiming to attain a better comprehension through interest and challenge. This chapter discusses an inductive process for finding the sum of interior and exterior angles of convex polygons by means of paper and scissors. The paper and scissors constitute auxiliary instruments for illustrating the concepts but also serve as scaffolds for building the inductive process, leading up to generalization.

The first polygon whose sum of angles is discussed is the triangle. Primary school teachers usually tell their pupils: "The sum of angles in a triangle is 180°". Using this phrase, we can deduce that the sum of angles of a square[1] is 360° as a polygon with four sides consists in fact of two triangles.

At the second stage, during the transition to polygons whose number of sides is greater than four, it is recommended presenting an inductive process, namely inference from the particular to the general. In our case, this implies generalization for finding the sum of interior angles of polygons. The conclusion drawn at the end of this process is that the greater the number of the polygon sides, the greater the sum of its angles. This conclusion applies to all polygon types, both concave and convex.

A similar process is performed when one attempts to find the number of diagonals in polygons. The first polygon which is normally presented to the learners is a quadrilateral and it is easy to demonstrate that a convex quadrilateral has only two diagonals. Thereafter, in an inductive process, we achieve generalization of finding the number of diagonals in a polygon with **n** sides. This generalization is valid to any type of polygon, both convex and concave. These two processes highlight the fact that when the number of polygon sides is greater, the sum of the interior angles is greater and so is the number of diagonals.

When teaching the topic of polygons, it is essential to distinguish between convex polygons and concave polygons, indicating to the learners

---

[1]Whenever the concept 'sum of angles' is used in this chapter, it implies the sum of **interior** *angles*. In the case of the sum of exterior angles, this will be explicitly indicated.

also the concept 'exterior angle'[2]. Before exploring the sum of exterior angles of a convex polygon, we should point out to the learners that the sum of exterior angles is relevant only to convex polygons and not to concave polygons. The objective is preventing learners from having the wrong concepts and committing common errors.

Based on the two previous inductive processes, namely finding the sum of interior angles of polygons or finding the number of diagonals, a similar assumption is likely to be made also in the case of the sum of the polygon exterior angles. That is, the greater the number of the polygon sides, the greater the sum of exterior angles. However, this is not the case. By making appropriate drawing and cutting, learners can reach the conclusive and surprising answer that there is no relation between the number of sides and the sum of exterior angles of any convex polygon. The sum is a determined size of 360°.

The Visual Representation in Geometry Teaching

Before presenting the inductive process of calculating the sum of interior and exterior angles of a convex polygon, it is only appropriate to discuss first the importance of visual representations in geometry teaching.

From its first steps, learning geometry involves visual representation (visualization). This is grounded in the fact that visual ability is required already at the first level of thinking according to levels theory conceived by Van Hiele (1987). No wonder then that learners whose visual ability is

[2]Exterior angle of a polygon is an angle formed by one side and by the continuation of the second side (Patkin & Levenberg, 2010).

not sufficiently developed, encounter difficulties in geometry learning and need to develop this ability as a prerequisite for their success later on. While teaching, we have to apply visual representations of all kinds, such as pictures, illustration aids, presentations, and movies. These items show geometry as ‘a real exhibit’ in the learners’ environment, serving as a bridge from the tangible to the abstract. They capture learners’ visual or tactile sense, enhance awareness of aesthetics and constitute an auxiliary instrument for understanding the functions of geometry in our life. Consequently, throughout the teaching process, we should encourage as much as possible the extensive use of visual representations. It is noteworthy that the special preparation of aids is not necessarily required. Some of these aids can be prepared together with the pupils. A wide variety of items around us can serve as illustration means for the learned topic, as is demonstrated in this chapter. Even the use of simple and available aids, e.g. paper and scissors, can be the beginning of a lesson with a fascinating visual representation. It all depends on the imagination and creativity level of teachers who wish to evoke interest and improve their pupils’ comprehension and attainments.

Drawing and cutting for the purpose of illustration and fostering pupils’ creativity

The use of illustration aids is one of the instruments designed for clear teaching of contents. As recommended by Eckhoff and Urbach (2008), this use might even encourage and support creative thinking.

A popular Chinese proverb says: “When I see, I identify, when I hear, I remember, when I do, I understand”. Patkin (1990) cites Provest (1985) and argues that learners should be given activities of 'doing geometry', such as drawing and inquiry for the purpose of enhancing their comprehension of the learned concepts. Torrance (1963) maintains that enabling learners to experience, play, inquire and ask questions is a good platform for developing their creativity. Similarly, Westcott (1978) stipulates that creative teaching of mathematics offers both learners and teachers an opening for exploring a variety of problem-solution ways and methods.

Cohen (2007) stipulates that "current use of relevant means designed for learners' active occupations originates in the faith that learning geometry is not a passive 'reading' of objects' attributes, but rather learning based on activities performed on them" (Cohen, 2007, p. 53).

Using paper and cutting with relation to polygons is not unfamiliar to geometry teaching practitioners (Heritage, 1966; Mountwitten, 2004; Patkin & Kanner, 2010). Patkin and Levenberg (2014) stipulate that paper folding facilitates such active experience, as creating straight lines in this way is an interesting way of finding and presenting relations between straight lines and angles. At a later stage, after experiencing and discovering new facts by means of paper folding, learners become acquainted with the formal use.

Studies show that educators' creative teaching enhances the development of their pupils' creativity (Lucas, 2001). Furthermore, it greatly promotes the teaching ability and professional development of the teachers themselves (Patkin, and Levenberg, 2017). According to Shriki (2015), such teaching style provides time and space to learners, encourages them to ask questions, turning them into active learners who assume responsibility for their own learning.

Using drawing and cutting as a simple and available illustration means, can make a considerable contribution to the learning of new concepts. Levenberg and Patkin (2014) argue that, following active experience with an inductive process, learners will be able to make a generalization about the sum of polygon angles.

To sum up, the didactic and pedagogical function of the assignment presented below aims to develop mathematics learning, based on experience and inquisitiveness as well as to provide a creative, visual and interesting mathematical activity.

Following is our recommendation for the stages in which the activity should be performed, as well as for the contents learned during it.

Calculating the sum of polygon angles by paper and scissors

Sum of triangle angles

1. First, a preliminary activity of experiencing with paper and scissors should be performed, in order to find out the sum of angles in a triangle. This topic is also included in a similar way in most textbooks.

2. Some triangle should be drawn and it should be freely torn into three parts. Then, the three angles should be connected as shown in Figure No. 2.

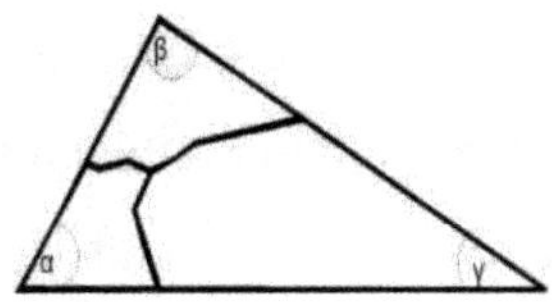

Figure No. 1

3. The two angles α and β should be placed next to angle γ. This visual illustration displays the sum of 180° (as in Figure No. 2).
4. This operation should be repeated with different types of triangles (e.g. obtuse-angled triangle and right-angled triangle) and with different side lengths. It is recommended performing the same operation about other angles, namely place angles α and γ next to angle β.

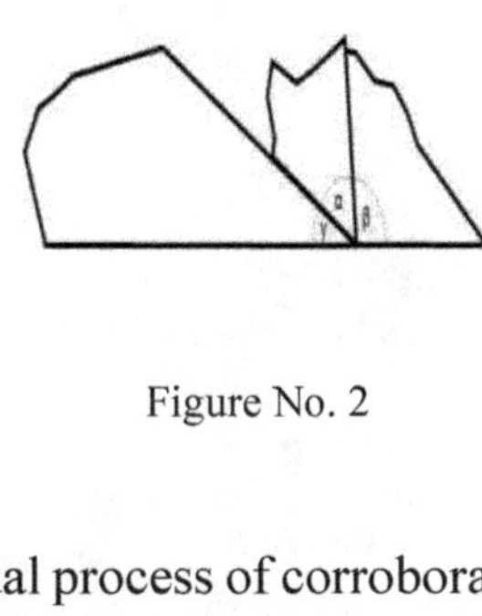

Figure No. 2

Thus, we present the visual process of corroborating the theorem: "the sum of angles in every triangle is 180°".

Sum of angles of any quadrilateral

1. A quadrilateral of any type should be drawn on an adhesive paper[3].
2. Four pieces of paper which include the angle should be cut in a free manner (i.e. without instructions) and arranged about a common vertex, as illustrated in Figure No. 3. (The cut angles are interior angles of the quadrilateral). By using the adhesive paper, we can determine that the sum of the quadrilateral angles is 360°.

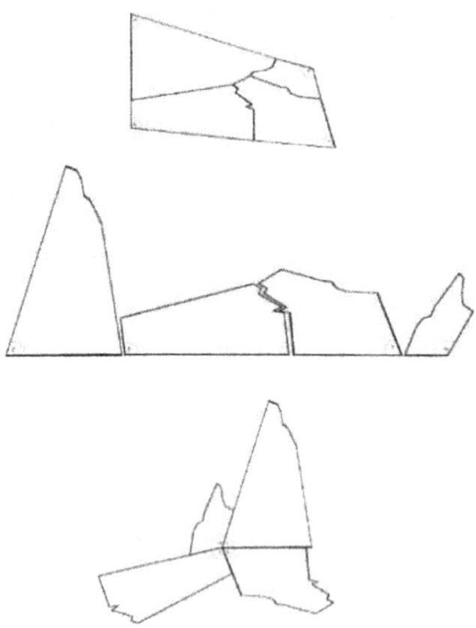

Figure No. 3

3. Through an inquiry activity, we have to assume and check whether arranging the angles in another order will result in another sum.

[3]Adhesive paper – a sheet of paper with a sticker on its other side. Removing the sticker uncovers glue which makes it possible to attach and glue the paper cuttings in the desirable location and sequence.

4. As a challenging assignment, we can check whether the sum of concave quadrilateral angles is also 360°. In that case, gluing the angle which is bigger than 180°must be done in an informed way.

Sum of angles of any pentagon

1. Some pentagon should be drawn on an adhesive paper and then cut.

2. Four pieces of paper that contain the angle should be cut in a free manner and arranged about a common vertex, similarly to the way we did it in the previous activity with the quadrilateral[4].

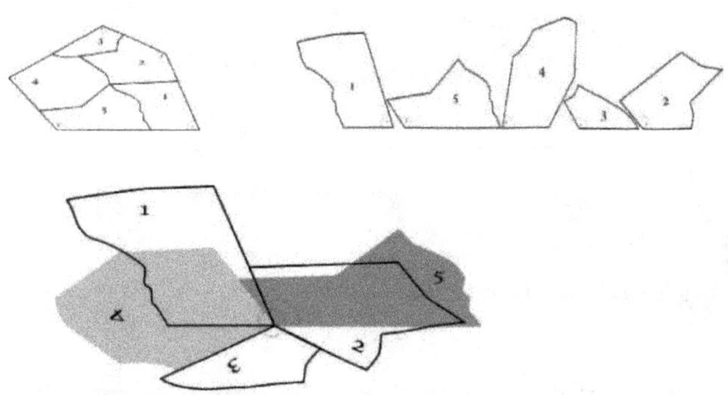

Figure No. 4a

3. During the inquiry activity, we have to assume and check whether arranging the angles in another order will result in another sum. Moreover, a challenging assignment can be suggested, namely

[4]During the process of adhesion, we are going to find out that the angles amount to more than 360° and we have to continue gluing. Some of the angles are bigger than the angles in the first circle and a circle and a half is obtained. That is, after completion to 360°, we are going to get additional 180°. Hence, the sum of a pentagon angles is 540°.

checking whether the sum of angles of a concave pentagon is equal to that of a convex pentagon.

4. A pentagon similar to the one we have drawn before should be drawn. The pentagon should be cut off the paper and then cut into a quadrilateral and a triangle.
5. The angles of the quadrilateral and of the triangle should be cut. Then all the quadrilateral angles should be positioned about a joint vertex. The same should be performed with the triangle angles[5]. In that way, we can determine that the sum of the pentagon angles is 540° (360°+180°).

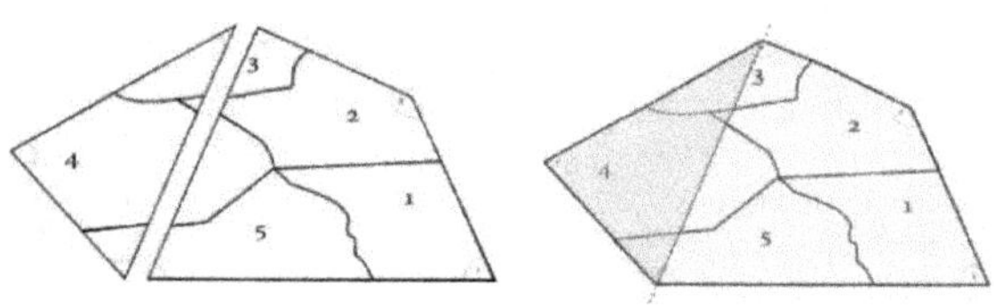

Figure No. 4b

Sum of angles of any hexagon

1. A hexagon of some type should be drawn on an adhesive paper and then cut.
2. Some pieces of paper which contain the angle should be cut in a free manner and arranged about a common vertex, similarly to the way it was done with the quadrilateral and the pentagon[6].

[5]In this way too we can determine that the sum of the pentagon angles is 540° (360°+180°).

[6]During the process of adhesion. we are going to find out that the angles amount to more than 360° and we have to continue gluing, until we have completed in fact two rounds, namely an addition of 360°. Hence, the sum of the hexagon angles is 720°.

3. Two pieces of paper should be taken and cut into three parts as illustrated in Figure No. 5. We have to make sure that only obtuse angles are obtained following the cutting process.

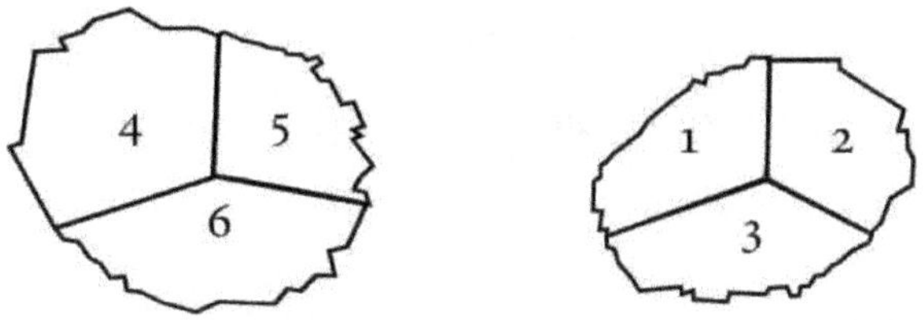

Figure No. 5

4. Each of the angles should be numbered and arranged according to the order displayed in Figure No. 6.

5. The resulting polygon is a hexagon.

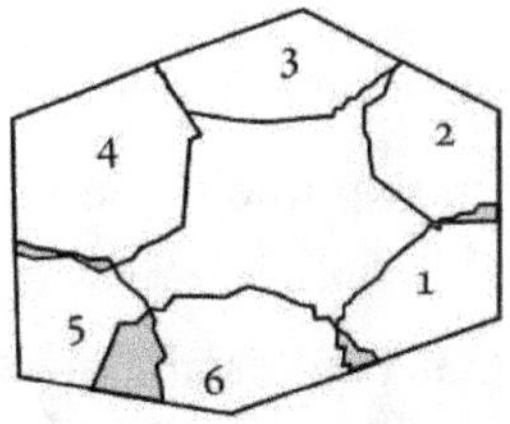

Figure No. 6

6. We have to check whether arranging the angles in another order is going to result in a different polygon.

7. As a challenging assignment, it is recommended checking which shape is obtained if there are five obtuse angles and one acute angle.

8. Another hexagon, similar to the one portrayed in Figure No. 7, should be drawn. (The hexagon can have whichever size you wish and it does not have to be a regular hexagon.)

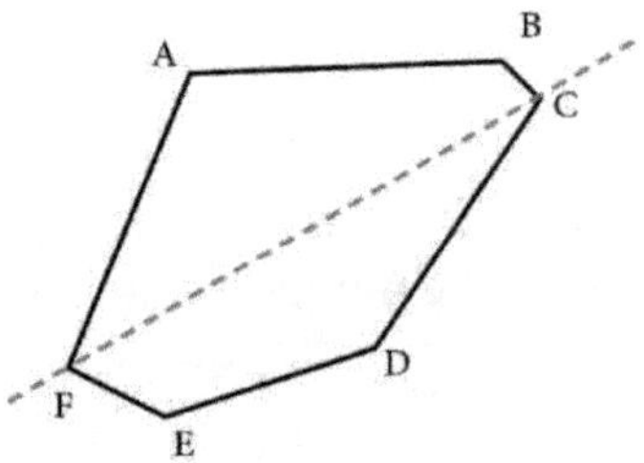

Figure No. 7

9. The hexagon should be cut out of the paper and later be cut it into two quadrilaterals (see the cutting line in the figure).

10. Four pieces of paper that contain the angle should be cut in a free manner. Then, all the angles should be placed about a common vertex. The obtained sum of angles should be examined.

11. This activity should be performed in more than one way.

Sum of a convex polygon angles

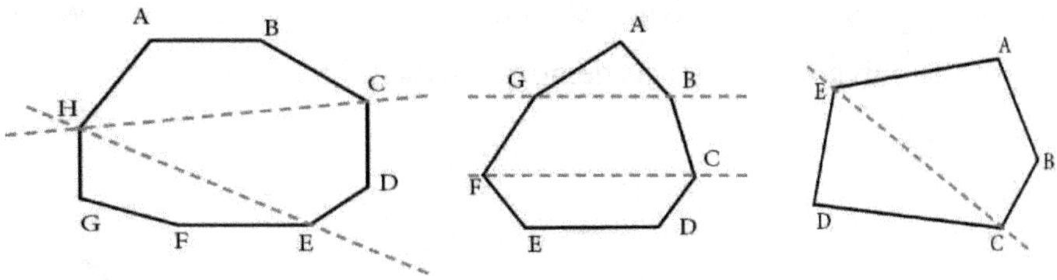

Figure No. 8

1. We have to name the polygons displayed in Figure No. 8 and estimate what would be the sum of angles in each of them.

2. The polygons in the figure are divided into quadrilaterals and triangles. Can the sum of the interior angles in each of them be calculated, without actually cutting each of the shapes? The sum of the interior angles in each polygon should be checked. This should be demonstrated by a drawing and the reasons for doing it should be explained. (Can we find this out only on the basis of the sum of the triangle angles?)

3. The sum of the interior angles of a 9-sided polygon and a 12-sided polygon should be calculated by using any way you choose.

Sum of convex polygon angles – using another way

1. A polygon with more than 4 sides should be drawn on the adhesive paper.

2. The polygon should be cut.

3. Some point on the polygon should be chosen and connected to each of the polygon vertices, as demonstrated in Figure No. 9.

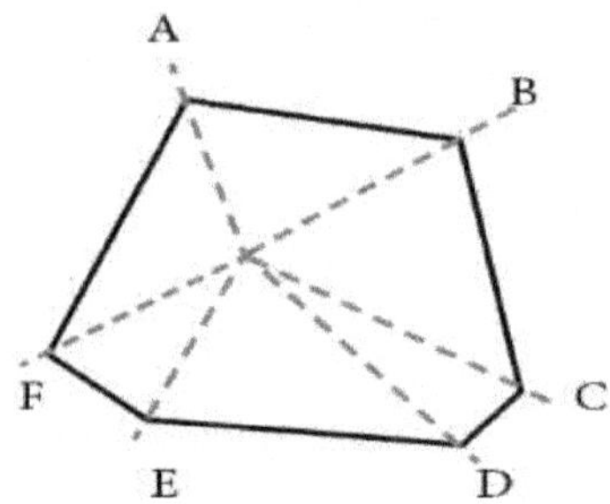

Figure No. 9

4. The angles about the point should be numbered and the polygon should be cut into triangles.
5. We have to check how many sides the polygon has and how many triangles are obtained?
6. We have to check the sum of angles in all the cut triangles as well as the sum of angles about - the point.
7. The sum of the other, non-numbered angles of the polygon should be calculated.
8. The sum of polygon interior angles should be calculated.
9. A list specifying all the sums of interior angles of the polygon examined in this activity should be prepared (the sum of interior angles should be indicated next to the number of the sides).
10. Can we know, based on the number of the polygon sides, also the sum of its interior angles?

Complete the drawing – to an n-sided polygon

Figure No. 10 shows the beginning of an incomplete polygon.

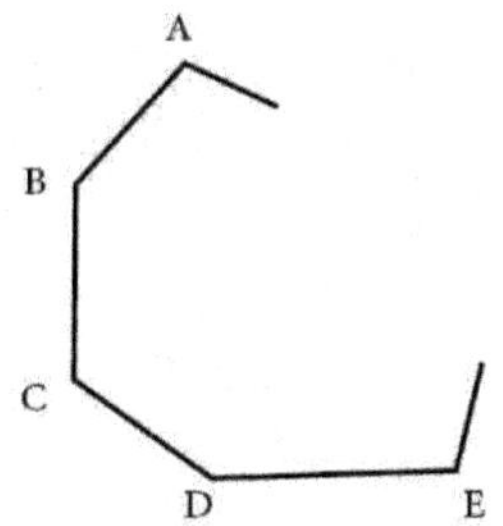

Figure No. 10

We are unaware of the number of the sides of this polygon and, hence, we cannot complete the drawing.

Assuming it has **n** sides, please answer the following questions:

a) Into how many triangles will the polygon be divided?
b) What is the sum of angles of all the obtained triangles?
c) What is the sum of angles about a point within the polygon? (Similar to what we did in example 10.)
d) What is be the sum of the other angles?
e) Using the formula for calculating the sum of interior angles of the polygon[7], please calculate the sum of interior angles of a quadrilateral, octagon, decagon, 12-sided polygon, 20-sided polygon, and 72-sided polygon.

[7]The formula for calculating the sum of interior angles of a polygon is (n-2)180°.

Sum of exterior angles in convex polygons.

Figure No. 11 shows a pentagon drawn on a piece of paper.

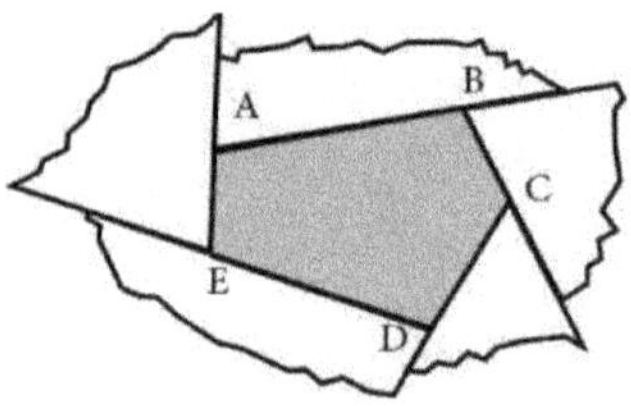

Figure No. 11

1. Cuts should be made along the straight lines.

2. You have to 'extract' the parts, as indicated in Figure No. 11 and arrange them as illustrated in Figure No. 12.

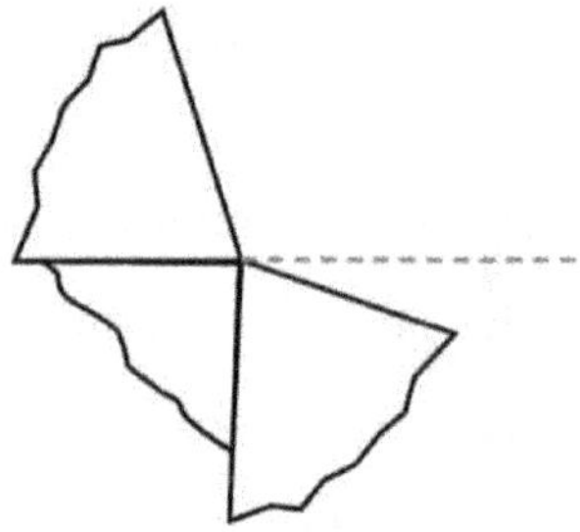

Figure No. 12

3. The performed activity is actually cutting off a polygon exterior angles. The obtained sum should be checked. The sum is surprising!

4. You have to check whether another sum is obtained if we change the order of the exterior angles' location. This should be done in the same way: make cuts along the straight lines, 'extract' the parts and arrange them.

Recommendation for further activity

1. Figure No. 13 presents different polygons. You have to indicate what is special about these polygons and what makes them different from the previous polygons in which we have engaged previously.

2. You have to calculate the sum of the interior angles of each of these polygons.

3. You should estimate whether it is necessary to cut and arrange angles as you did in the previous assignment, i.e. division into triangles, or the sum of the interior angles of each polygon can be calculated without cutting.

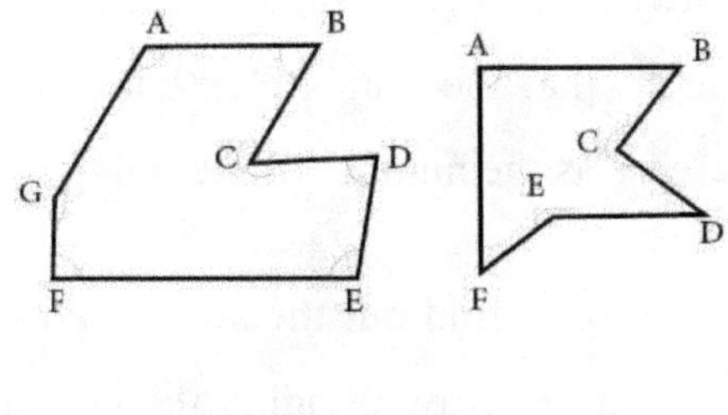

Figure No. 13

Difficulties and anticipated errors

Those engaged in the activity suggested above should be aware of common and prevalent errors that learners commit when calculating the sum of polygon angles.

a. One mistake is the intuitive assumption that the sum of exterior angles in a concave polygon increases as the number of polygon sides increases. In this case, the intuition according to which the pattern will be identical to the pattern applicable to the sum of a polygon interior angles or the number of diagonals in a polygon is extremely misleading. In their study, Jung, Kloorsterman, and McMuller (2007) explained this error. According to them, young learners view models intuitively, beyond what is expected of them. They arrive at problem-solution strategies by means of the same models, while relying on the same capability which is developed in them in a natural manner.

b. Another error is reversed thinking. If an exterior angle of a convex polygon complements the interior angle to the sum of 180°, then this leads to the thought that the sum of exterior angles in a convex polygon will decrease as the number of the polygon sides increases.

c. One way of attempting to find out the sum of polygon angles (also mentioned in this chapter) is by dividing the polygon into triangles. This is not necessarily done by diagonals, but rather by indicating an interior point within the polygon and creating triangles by drawing segments which connect the triangle vertices with that internal point, similar to (Section f, Figure No. 9). If we count the triangles obtained in the drawing and multiply by 180°, we will get a wrong answer for the sum of the triangle angles. In this case, the resulting sum is higher by 360° than the correct sum because the wrong sum includes also the sum of all the angles formed about the inner point. At this time, learners should be made aware of the source of the error. Performing a process of control and re-examination of those angles whose sum we are trying to calculate, improves the learning process and contributes to deep meaningful learning.

Summary

The activities included in this chapter apply to paper and scissors for illustrating, stage after stage in an inductive way, how to find a formula for calculating the sum of interior and exterior angles in convex polygons. The goal was to present a rich and in-depth inquiry assignment which integrates discovery, assumptions, validation, proving processes, explanation, asking questions and giving reasons. It is recommended using this approach also while teaching other topics of geometry. This way of acting is important not only to primary school pupils but also to older students and even to pre-service mathematics teachers. Experiencing with inductive processes, combined with inquiry processes already at a young age, greatly promotes the development of mathematical thinking. Integrating illustration and calculation in numbers leads to generalization and constitutes an example for developing geometric thinking and nurturing creativity. Every experience of this kind, directly and indirectly, enhances mathematical insight and improves attainments in this field.

References

1. Cohen, N. (2007). *Analytical-Visual Integration in 3-D. An analysis of prospective teachers and teachers thinking processes.* Doctoal dissertation. Beer-Sheva: Ben-Gurion University of the Negev. [Hebrew]
2. Eckhoff, A., & Urbach, J. (2008). Understanding imaginative thinking during childhood: Sociocultural conceptions of creativity and imaginative thought. *Early Childhood Education Journal, 36,* 179–185.
3. Heritage, R. S. (1966). *Learning Mathematics, Book one* (pp. 199–205). Harmondsworth, Middlesex, England: Penguin Books Ltd.
4. Jung, M., Kloosterman, P., & McMullen, M. B. (2007). Young Children's Intuition for Solving Problems in Mathematics, *Young Children, 62*(5), 50.
5. Levenberg, I., & Patkin, D. (2014). Investigating the Sum of the Interior Angles of Convex Polygons using Paper and Scissors. *Workshop in: Children's Mathematical Education, CME, 2014, Poland,* pp. 35–36.
6. Lucas, B. (2001). Creative teaching, teaching creativity and creative learning. In A. Craft, B. Jeffrey, & M. Leibling (Eds.), *Creativity in education* (pp. 35–44). London: Continuum.
7. Mountwitten, M. (2004). Disassembling and assembling of polygons. *Mispar Hazak, 8*, 34–37. [Hebrew]
8. Patkin, D. (1990). *The utilization of computers: Its influence on individualized learning, pair versus individualistic learning. On the perception and comprehension of concepts in Euclidean geometry at various cognitive levels within high school students.* Doctoal dissertation. Tel Aviv: Tel-Aviv University. [Hebrew]
9. Patkin, D., & Kanner, B. (2010). Special Segments of Triangles, *The Mathematics Educator, 12*(2), 25–50.
10. Patkin, D., & Levenberg, I. (2010). *Solid geometry – Part I for junior high and high school,* (2nd ed.) (p. 104). Published by the authors. [Hebrew]
11. Patkin, D., & Levenberg, I. (2014). Investigating the Sum of Interior Angles of Convex Polygons using paper and scissors. *Learning and Teaching Mathematics. A Journal of AMESA, 16,* 7–12. *http://frink.machighway.com/~dynamicm/LTM16-complete.pdf*

12. Patkin, D., & Levenberg, I. (2017). Co-Teaching during an in-service training course in geometry. In D. Jarvis & M. Kariuki (Eds.), Co-Teaching in *Higher Education: Creative Approaches to Co-Planning, Co-Instruction, and Co-Assessment in Practice and Research.* [31 pages] Canada: University of Toronto Press.
13. Pedersen, J. Retrieved August 2013 from: http://www.onlinemathlearning.com/geometry-math-quotes.html
14. Provest, F. J. (1985). Geometry in the junior high school. *Mathematics Teacher, 78*, 411–418.
15. Shriki, A. (2015). Creativity has many faces. In: A. Gazit & D. Patkin (Eds.), *Mathematical thinking and creativity in the solution of challenge problems in mathematics, strategies, dilemmas and mistakes*.
16. Torrance, E. P. (1963). Creativity. in F. W. Hubbard (Ed.), *What research says to the teacher, 28*, Washington, DC: Department of Classroom Teachers American Educational Research Association of the National Education Association.
17. Van Hiele, P. M. (1987). *Van-Hiele levels: A method to facilitate the finding of levels of thinking in geometry by using the levels in arithmetic*. Paper presented at the conference on learning and teaching geometry: Issues for research and practice. Syracuse University.
18. Westcott, A. M. (1978). *Creative teaching of mathematics in the elementary school*. Boston, MA: Allyn and Bacon.

https://doi.org/10.1142/9789811218750_0004

Chapter 4

On areas, perimeters, and creativity

Ronit Bassan-Cincinatus

Creativity is a dynamic, growing and developing a process whereby we take the existing condition, re-shape it and form new connections in order to solve a certain problem (Grossman, Rogers & Mor, 1995) or to promote and upgrade an existing situation (De Bono, 1993). Fostering creative ability is a prerequisite in a rapidly changing world and in a dynamic education system for both teachers and learners. A high creative ability will assist in improving the teaching and in enhancing the level of academic attainments.

The actual need for creativity exists wherever there is a real need for a new idea. Creativity helps in the case of a problem or a conflict or when other approaches failed and a new idea might open an opportunity and it can bring about advantage or benefit. Whenever thinking is required, there is also a need for information, analysis, and creativity. Every creative practice involves a certain manipulative action leading to the creation of new links or connections to familiar situations. Creative practice leads to an idea, production or model, which is sometimes new and sometimes improved and more sophisticated (Landau, 1973).

Creativity in mathematics enables the presentation of problems that have no immediate and visible solution or of problems that have several solutions or various solution methods. Inquiry activity or directed inquiry entails situations for creative thinking which embody unexpected discoveries or new inquiry directions (Yee, 2005). Most mathematics teachers acknowledge the importance of their pupils' development (Shriki, 2010). Nevertheless, the research literature illustrates that usually, pupils do not have the opportunity to study in learning environments that foster their mathematical creativity (Sriraman, 2005). Both the major part of the curriculum and the pedagogical approaches at school hardly allow pupils to encounter the open facet of mathematics, avoiding opportunities to present open problems to the pupils.

Geometry and especially the Euclidean geometry evokes academic situations that facilitate encouragement and development of creative thinking. One of the goals of geometry studies at elementary school is fostering geometrical insight (Ministry of Education, Culture and Sport, 2006). This ability comprises, among others, an ability to explore the shapes themselves, their properties and interrelations. In addition to areas and perimeters calculation assignments, elementary school pupils should cope with assignments that require the development of thinking and understanding, allow the development of visual and spatial perception as well as an in-depth inquiry of basic and complex shapes.

A good way of increasing pupils' learning motivation is a studying environment that offers experiential learning. Thus, learning is done in an active rather than passive manner. The assignments in this chapter enable pupils to construct knowledge, perform analysis and synthesis, be creative in order to present the way of thinking in addition to the way of arithmetic calculation as well as to draw conclusions.

Bloom's taxonomy (cited in Cook, 2010) presents six categories of cognitive goals which learners are required to accomplish while executing tasks that enable effective teaching and learning: knowledge, comprehension, application, analysis, synthesis, and assessment. The element of creativity is not included in a direct manner but it is manifested at high levels of synthesis and assessment. Anderson (cited in Cook, 2010), Bloom's disciple, apparently maintained that the element of creativity should be overt in the taxonomy. Hence she introduced a certain change in the categories: remembering, understanding, applying, analyzing, evaluating and creating. The first category was changed from knowledge to remembering, the next three categories remained identical to Bloom's categories. Then the element of synthesis was replaced by evaluating which was ranked fifth, followed by creating in the sixth place. Taking into consideration that one of the leading definitions views creativity as an ability to create a new essence by means of existing elements (Nevo, 1997), then a synthesis combined with evaluation in Bloom's taxonomy manifests the element of creativity in Anderson's taxonomy.

It is noteworthy that not all the levels should be used in the development of activities; this depends on the learning material and on the goals. Below is a demonstration of the way the different levels are manifested in the area and perimeter assignments:

Remembering: this level requires remembering and identification of the shape elements, shape reconstruction, namely a goal that can be accomplished by remembering only, without a need for a higher level.

Understanding: a level of thinking which can ascertain the ability to use the knowledge – translation, interpretation, and extrapolation. For example, when we take four quarters of a circle which are interconnected in a different way (perhaps even on a different scale), the understanding facilitates finding the similarity between them and the whole circle.

Applying: the ability to use principles, generalizations and ideas in situations which are new for the pupils (thus, their extent of understanding the idea is manifested). For example, if we cut a circle into four quarters, the area of the four quarters is always the whole, regardless of the way the quarters are constructed.

Analyzing: the ability to decompose the assignment or the complex shape into basic elements so that the relationship between the parts of the complex shape becomes clear. For example, the distinction between the relevant (the gray, highlighted) area and the irrelevant (the white, not highlighted) area.

Evaluating: the ability to judge (qualitatively or quantitatively) the values of ideas, solutions and so on while using criteria or standards in order to determine accuracy, effectiveness, advantages, disadvantages, etc.

Creating: the ability to create (in a qualitative or quantitative way) new shapes and new ideas from new points of view based on acquired previous knowledge.

At school, most lessons are dedicated to knowledge, comprehension and some application. However today, following the new reforms, verbalization of thinking is nurtured in order to develop the ability to think which comprises elements of creative thinking. This is a key principle in the explicit teaching of meta-cognition and all its components. Most studies show that meta-cognitive approaches improve not only learners' ways of thinking but also their attainments (Eden, Mevarech & Koretz, 2008).

The principle of the spoken language is grounded in the theory conceived by Vygotsky (1978). This theory advocates the importance of social interactions and uses of the spoken language as a "tool" for attributing meaning to knowledge and developing the thinking ability. Verbalization of the thinking should be implemented, orally or in writing, in every appropriate opportunity. During lessons in class verbalization of thinking can be used orally by thinking aloud and creating a language of thinking which is shared and common. This can be done by the teachers' verbalization in front of the pupils (modeling) and by requesting the pupils to verbalize their thinking within the framework of directed class/group discussions.

The teaching of creative thinking which is conscious, controlled and focused on specific contents, improves creative thinking in general, creative thinking in science and in scientific thinking. Aran & Ashkenazi (2008) studied the field of sciences teaching. They found that pupils exposed during their studies to activities and assignments which develop their creative thinking, demonstrate a higher level of inquiry and question asking competences. This is manifested by asking more questions that are of various types and in varied areas. Consequently, the thinking skills and characteristics that might stimulate creativity should be fostered as much as possible among pre-service teachers, in-service teachers, and learners.

This chapter aims to develop and enhance the competences of the area and perimeter calculations while applying cognitive abilities. The combination of two familiar basic shapes – a square and a circle/sphere – leads to the formation of complex shapes that can be decomposed in various ways. Some of the learning topics discussed in this chapter are: calculating the area and the partial and/or full perimeter of the geometric shapes square and a circle/sphere [hereunder – "the basic shapes"]; calculating the area and perimeter of new geometric shapes, namely the result of combining the "basic shapes" from which they are created [hereunder – "the complex shapes"].

Parallel to the learning goals of enhancing the abilities of area and circumference calculations, this paper discusses additional competencies that can promote the decomposition and composition skills of the basic and complex shapes. They can also enhance the learners' ability to form new shapes following the activities and the acquired knowledge. Moreover, the assignments presented below develop pupils' visual perception and their ability to calculate the area and perimeter of a new geometrical shape obtained from the integration and synthesis of the basic geometrical shapes. This ability develops skills of identification and understanding the meaning of the missing part and its completion to a whole, based on the area and circumference of the basic geometrical shapes familiar to the pupils. An activity of this type facilitates both individual and collaborative learning. Individual learning consists of assignments which are personal rather than collective and therefore they stimulate a personalized and independent process of thinking and learning. Collaborative learning transpires when teachers divide the pupils into groups (preferably small groups of two or three members) by taking into consideration various features (e.g. mathematical abilities, social skills). When the groups are heterogeneous, the pupils can mutually contribute to and benefit from one another according to their strong and weak points. Thus, for example, a pupil who is endowed with better capabilities in algebra can be paired with pupils whose capabilities in geometry are higher.

The following activities allow the structuring of knowledge. The shape sequence is given so that from one complex shape to another, learners can draw conclusions and achieve high levels of understanding. The complex shapes are organized in a certain sequence, enabling the implementation and application of previous ideas for the purpose of consolidating ideas and generalizations about new situations. Decomposing the complex shape into fundamental elements, exposes and clarifies the relationship between parts of the complex shape. Learners' abilities to understand, decompose and compose the shapes into new shapes or to create a completely new sequence of assignments, develop their visual perception and creativity.

Below are four activities of calculating the areas and perimeters of complex shapes:

Activity 1:

This activity is divided into two parts: in the first part one has to calculate the area of all the blackened shapes and in the second part their perimeter. All the shapes were obtained from basic shapes of a square and a sphere. To be more specific, each of these complex shapes was created from a square which circumscribed a sphere, several spheres or parts of a sphere. In all the shapes, the length of the square's side, which is identical in all the shapes, is given. According to it and to the drawing, the length of the circles' radius can be calculated.

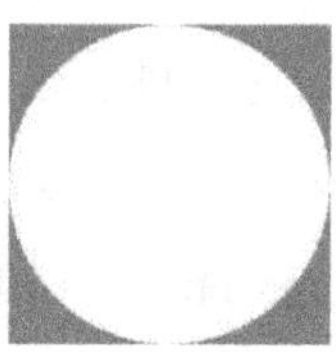
(1)

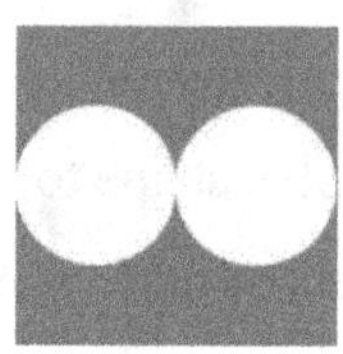
(2)

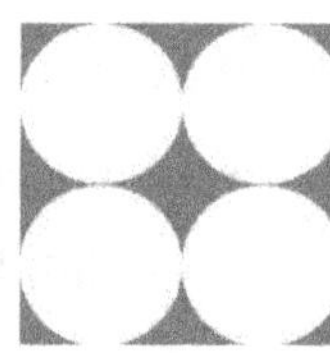
(3)

(4)

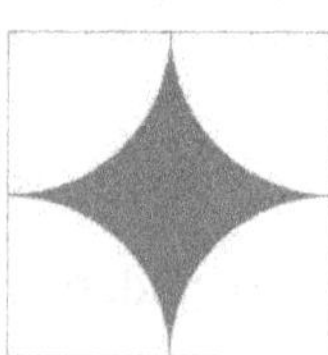
(5)

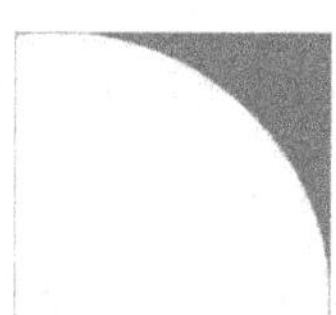
(6)

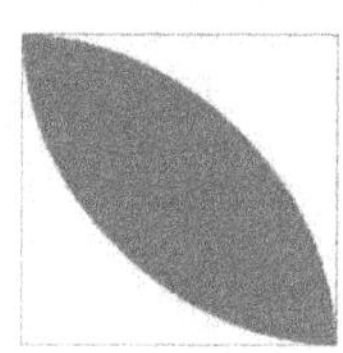
(7)

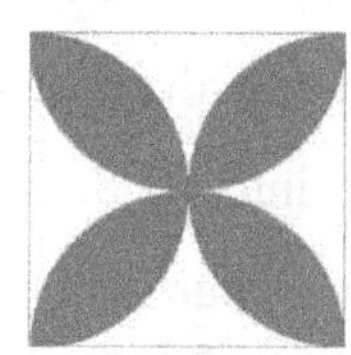
(8)

As mentioned, it is important to verbalize ways of thinking before writing the solution. Consequently, the ideas before calculating the area and perimeter of the blackened shapes were represented in words:

(1) The area is the area of the square minus the area of the sphere; the perimeter is the perimeter of the square plus the perimeter of the circle.

(2) The area is the area of the square minus the area of the two spheres; the perimeter is the perimeter of the square plus the perimeter of the two circles.

(3) The area is the area of the square minus the area of the four spheres; the perimeter is the perimeter of the square plus the perimeter of the four circles.

(4) The area is the area of the square minus the area of the three quarters the sphere; the perimeter is the perimeter of the square plus the perimeter of the three-quarters of the circle plus the two circle's radii (or length of the square's side).

(5) The area is the area of the square minus the area of the four quarters of the sphere, that is the sphere is cut out of the square. The perimeter is the perimeter of the four quarters of the circle, namely the circle.

(6) The area is the area of the square minus the area of the one-quarter of the sphere; the perimeter is the perimeter of half the square plus the perimeter of one-quarter of the circle.

(7) The area is the area of the square minus twice the area of shape 6; the perimeter is the perimeter of the two-quarters of the circle, namely the perimeter of half the circle.

(8) The area is four times the area of shape 7 and the perimeter is four times the perimeter of shape 7, the length of the square's side and the circle's radium being half the lengths in shape 7 (the square can be divided into four identical squares, each of them being a reduction of shape 7).

Activity 2:

The second activity is also divided into two parts: in the first part, one should find the area of all the blackened shapes and in the second part, their perimeter. These shapes resulted from the shapes of the first activity. In this activity too, all the shapes were created from basic shapes of a square and a sphere. These complex shapes were formed with reference to the shapes of the first activity. They are indeed similar to yet different from them, i.e. in some of them the colors were substituted, the black became white and vice versa. In these shapes too, the length of the square's side which is identical is given. According to the drawing, the length of the circle's radius can be calculated.

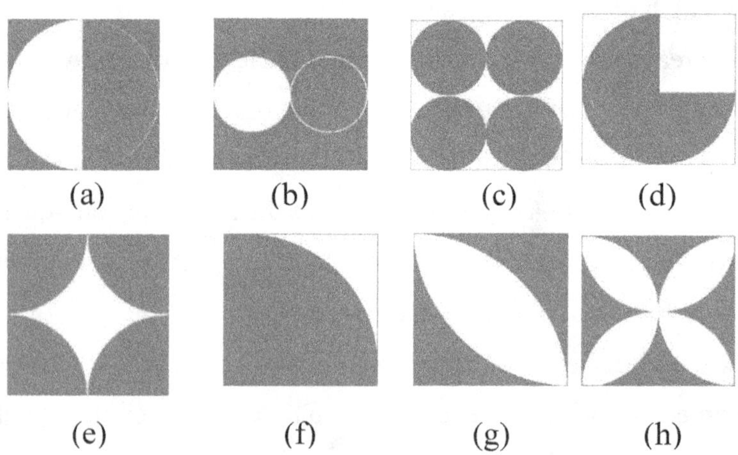

The areas and perimeters of the blackened shapes: it is important to see what remains white, what does not belong to the shape's area and should, therefore, be "cut out". Then, it becomes easier to draw the outlines of the shape and calculate its perimeter.

a) The area is the area of the square minus the area of half the sphere; the perimeter is the perimeter of the square plus the perimeter of half the circle plus the diameter's length (or the length of the square's side). The relation between shape 1 and shape (a) is visible.

b) The area is the area of the square minus the area of the sphere; the perimeter is the perimeter of the square plus the perimeter of the circle. The relation between shape 2 and shape (b) is visible.

c) The area is the area of the four spheres; the perimeter is the perimeter of the four circles. The connection between shape 3 and shape (c) is visible. However, in shape 3, one has to relate to the differences (the area of the square minus the area of the four spheres) whereas in shape (c) there is no need for area differences at all.

d) The area is the area of three-quarters of the sphere; the perimeter is the perimeter of the three-quarters of the circle plus the length of two radii (or the square's side). The connection between shape 4 and shape (d) is visible. In shape 4, one has to relate to the area difference (area of the square minus the area of three-quarters of the sphere) whereas in shape (d) there is no need for area differences at all.

e) The area is the area of the four-quarters of the sphere (the area of the sphere); the perimeter is the perimeter of the square plus the perimeter of the four quarters of the circle (the area of the circle). (for visual demonstration and explanation, the whole circle should be drawn around each quarter of the circle). The relation between shape 5 and shape (e) is visible. In shape 5, one has to understand what is the shape. It is important to see that one cuts off the square the four quarters of the sphere (area of the square minus area of the four quarters of the sphere) and in shape (e) one should cut shape 5 off the square.

f) The area is the area of one-quarter of the sphere; the perimeter is the perimeter of half the square plus the perimeter of one-quarter of the circle. The relation between shape 6 and shape (f) is visible. In shape 6, one should understand what is the shape. It is important to see that one cuts off the square the quarter of the sphere and in shape (f), one cuts shape 6 off the square.

g) The area is twice the area of the square minus the area of one-quarter of the sphere; the perimeter is the perimeter of the square plus twice the perimeter of one-quarter of the circle. The relation between shape (f) and shape (g) is visible. The white area in shape (f) has become black in shape (g). Another relation between shape 6 and shape (g) is visible. It was indicated that shape 6 was obtained from the area of the square minus the area of one-quarter of the sphere. Hence, shape (g) is twice the area of shape 6. Yet another relation can be noticed between shape 7 and shape (g). It was indicated that the areas of shape (g) and shape 7 create together the area of the square.

h) The area is four times the area of the shape (g), the square's side being half the square's side in shape (g) and the circle's radius is half the sphere's radius in shape (g). The perimeter is the perimeter of the square plus eight times one-quarter of the circle's perimeter whose radius is half the circle's radius in shape (g). The relation between shape 8 and shape (h) is visible. The white area in shape 8 has become black in shape (h). Another relation between shape 5 and shape (h) is visible: shape 5 appears twice in shape (h).

Activity 3:

The third activity can be just an activity of the area and perimeter calculations. However, this type of activity which is calculation-oriented only does not stimulate creativity at all. The purpose of the activity, beyond area and perimeter calculations, is to lead to comparisons between the areas of the two blackened shapes and the perimeters thereof. The feeling that the areas and perimeters are not equal is strong. Consequently, a creative solution will entail explanations, arguments and solving strategies that are varied and creative.

In order to enrich and enhance the experience, one can compare the area and perimeter of two additional complex shapes. These are identical shapes whereby the blackened part is the "complementary", the opposite, of the white part.

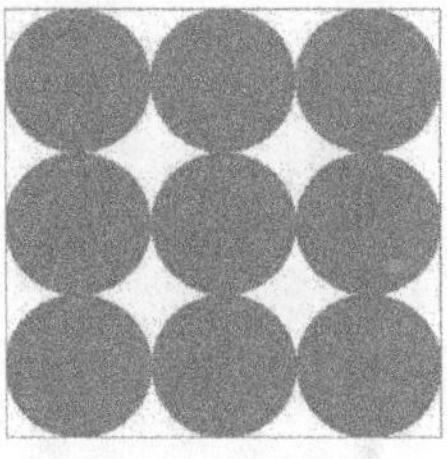

Activity 4:

In the fourth activity, the area and perimeter of the blackened shape in the drawing has to be calculated. The shape was formed from the basic shape of a sphere alone. There is, in fact, the "misleading" frame of the square which circumscribes the circle but this enables the radius to be calculated (the square is not included in the calculation of the area or the perimeter of the complex shape. It is drawn only for clarifying the length of the radius).

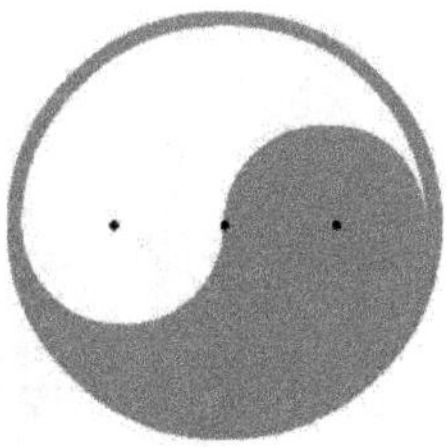

Solutions of Activity 1

1. **The side of the square: 20 cm,**

   the radius of the circle: $\frac{20}{2} = 10\ \text{cm}$

Area – area of the square minus the area of the sphere.

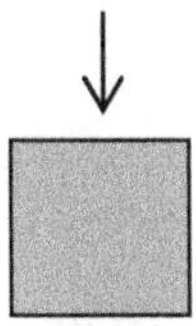

$20^2 = 400$

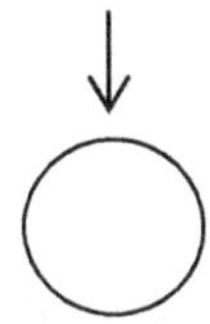

$\pi\left(\frac{20}{2}\right)^2 = 100\pi$

The total of the blackened area is the area difference between the square and the sphere. Consequently,

$$20^2 - \pi\left(\frac{20}{2}\right)^2 = 400 - 100\pi \text{ cm}^2$$

Perimeter – perimeter of the square plus circumference of the circle.

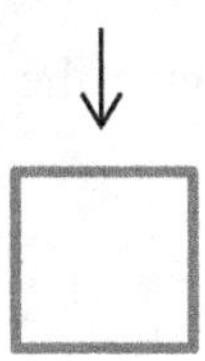

$4 \times 20 = 80$

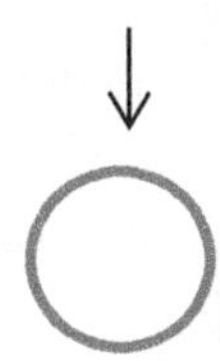

$2\pi \times \frac{20}{2} = 20\pi$

The total of the blackened shape's perimeter is the amount of the square and circle perimeters. Hence: $80 + 20\pi$ cm.

2. **Side of the square: 20 cm, radius of the circle:** $\frac{20}{4} = 5$ cm.

Area – the area of the square less twice the area of the sphere

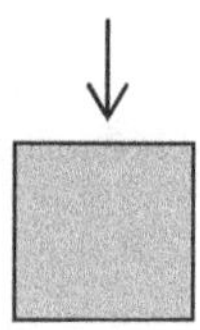

$20^2 = 400$

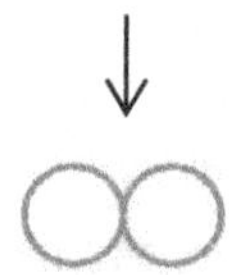

$2 \times \pi \times 5^2 = 50\pi$

The total of the blackened area is the area difference between the square and the two spheres. Consequently: $400 - 50\pi$ cm^2.

Perimeter: perimeter of the square plus twice the perimeter of the circle

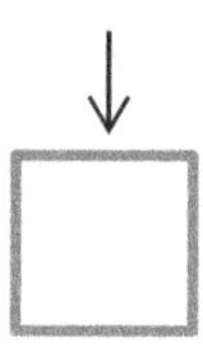

$4 \times 20 = 80$

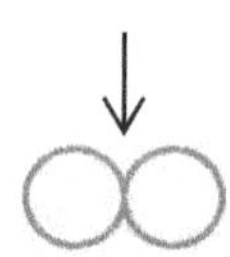

$2 \times 2\,\pi \times \frac{20}{4} = 4\pi \times 5 = 20\pi$

The total perimeter of the blackened shape is the amount of the square and two circles perimeters. Consequently

$$4 \times 20 + 2 \times \left(2\pi \frac{20}{4}\right) = 80 + 4\pi \times 5 = 80 + 20\pi \text{ cm.}$$

3. **The side of the square: 20 cm, the radius of the circle:** $\frac{20}{4}$ cm.

Area - the area of the square minus four times the area of the sphere

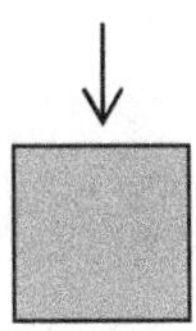

$20^2 = 400$

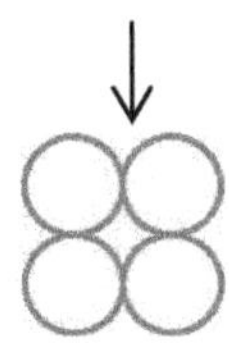

$4 \times (\pi \times (\frac{20}{4})^2 = 4\pi \times 5^2 = 100\pi$

The total of the blackened area is the area difference between the square and the four spheres. Consequently:

$$20^2 - 4\pi\left(\frac{20}{4}\right)^2 = 400 - 100\pi \text{ cm}^2.$$

Perimeter – the circumference of the square plus four times the circumference of the circle.

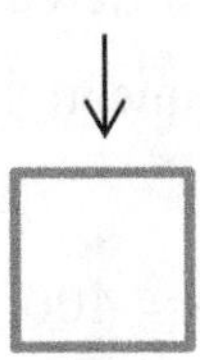

$4 \times 20 = 80$

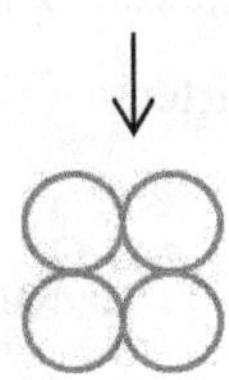

$4 \times (2\pi \times \frac{20}{4}) = 4 \times 2\pi \times 5 = 40\pi$

The total perimeter of the blackened shape is the amount of the square and the four circles perimeters. Consequently:

$$4 \times 20 + 4 \times \left(2\pi \frac{20}{4}\right) = 80 + 40\pi \text{ cm.}$$

4. The side of the square: 20 cm, the radius of the circle: $\dfrac{20}{2} = 10$ cm.

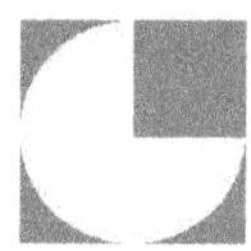

Area – area of the square minus three times the area of one-quarter of the sphere.

$$20^2 = 400$$

$$3 \times \frac{\left[\pi \left(\frac{20}{2}\right)^2\right]}{4} = 3 \times \frac{10^2\pi}{4} = 75\pi$$

The total of the blackened area is the area difference between the square and the three-quarters of the sphere. Consequently:

$$20^2 - 3 \times \frac{1}{4} \times \pi \left(\frac{20}{2}\right)^2 = 400 - \frac{3}{4} \times \pi \times 100 = 400 - \frac{75}{4} \times \pi \text{ cm}^2.$$

Perimeter – The square perimeter plus three-quarters perimeter of the circle

$4 \times 20 = 80$

$\frac{3}{4} \times \left(2 \times \pi \times \frac{20}{2}\right) = 15\pi$

plus two radii

$2 \times \frac{20}{2} = 20$

The total perimeter of the blackened shape is the sum of perimeters of the square, three-quarters of the circle and the two radii, namely perimeter of the square and perimeter of the segment. Consequently:

$$4 \times 20 + \frac{3}{4} \times 2\pi \frac{20}{2} + 2 \times \frac{20}{2} = 80 + 15\pi + 20 = 100 + 15\pi \text{ cm}.$$

5. **Side of the square: 20 cm, radius of the circle:** $\frac{20}{2} = 10$ **cm.**

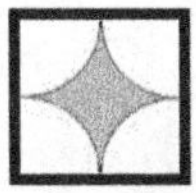

Area – area of the square minus four times the quarter of the sphere.

$20^2 = 400$

$4 \times \frac{\pi \left(\frac{20}{2}\right)^2}{4} = 100\pi$

The total blackened area is the area difference between the square and the four quarters of the sphere which are a circle. Consequently:

$$20^2 - 4 \times \frac{1}{4} \times \pi \left(\frac{20}{2}\right)^2 = 400 - \pi(10)^2 = 400 - 100\pi \text{ cm}^2.$$

Perimeter – the four quarters of the circle perimeter.

$$4 \times \frac{1}{4} \times \left(2\pi \times \frac{20}{2}\right) = 20\pi \text{ cm}$$

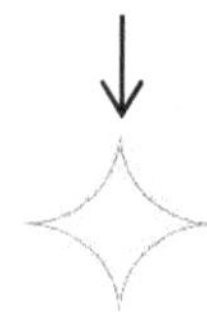

The total perimeter of the blackened shape is the perimeter of the circle.

6. The side of the square: 20 cm, the radius of the circle: 20 cm.

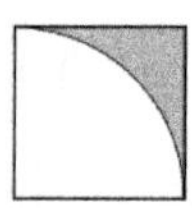

Area – the area of the square minus one-quarter of the sphere.

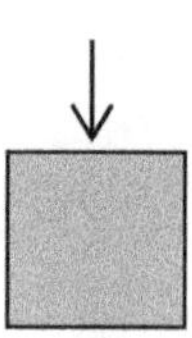

$$20^2 = 400$$

$$\frac{\pi \times 20^2}{4} = 100\pi$$

The total of the blackened area is the area difference between the square and one-quarter of the sphere. Consequently:

$$20^2 - \frac{\pi \times 20^2}{4} = 400 - 100\pi \text{ cm}^2.$$

Perimeter – the perimeter of half the square plus the perimeter of a quarter of the circle.

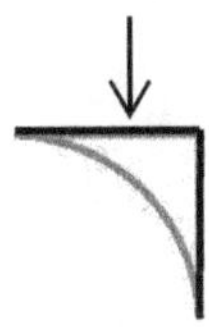

$\frac{1}{4} \times (2\pi \times 20) = 10\pi$

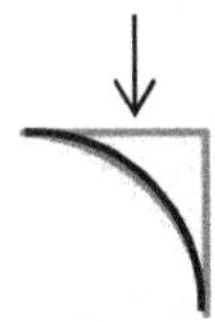

$4 \times 20 = 80$

The total of the blackened shapes' perimeter is the amount of the perimeter s of half the square and the perimeter of one-quarter of the circle. Consequently:

$$2 \times 20 + \frac{1}{4} \times 2\pi \times 20 = 40 + 10\pi \text{ cm}.$$

7. Side of the square: 20 cm, radius of the circle: 20 cm.

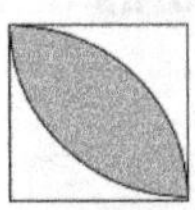

Area – the area of one-quarter of the sphere less the area of shape 6,

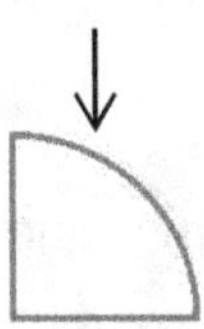

$\frac{\pi \times 20^2}{4} = 100\pi$

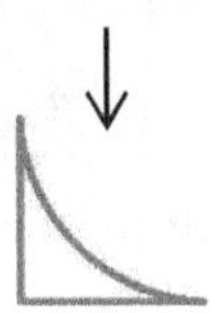

$400 - 100\pi$

The total area of the blackened area is the area difference between a quarter of the sphere and shape 6, Consequently:

$$\frac{\pi \times 20^2}{4} - (400 - 100\pi) = 100\pi - 400 + 100\pi = 200\pi - 400 \text{ cm}^2.$$

Perimeter – twice the perimeter of one-quarter of the circle.

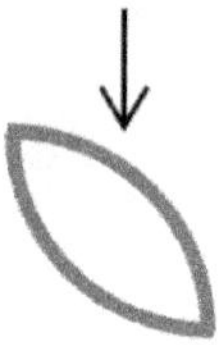

$$2 \times \left(\frac{1}{4} \times (2\pi \times 20)\right) = 20\pi$$

The total perimeter of the blackened shape is the amount of the two perimeters of the quarter of the circle, namely half the perimeter of a circle. Consequently: $2 \times \frac{1}{4} \times 2\pi \times 20 = 20\pi$ cm.

8. Side of the square: 20 cm, radius of the circle: $\frac{20}{2} = 10$ cm.

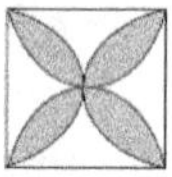

Area – Way A

One can look at the similarity between shape 7 and shape 8. Shape 7 appears four times in shape 8 but it is reduced. We will divide the square of shape 8 into four squares so that every square circumscribes the reduced shape 7. We will observe the reduced square which is a quarter of the large square:

Side of the reduced square: $\frac{20}{2} = 10$ cm, the radius of the reduced circle: $\frac{20}{2} = 10$ cm.

The total of the blackened area is four times shape 7. Hence, everywhere the length of the square's side is 20, we will write: $\frac{20}{2} = 10$ cm. Consequently:

$$4 \times \left(\frac{\pi \times 10^2}{4} - (100 - 25\pi) \right) = 4 \times (25\pi - 100 + 25\pi)$$
$$= 4 \times (-100 + 50\pi) = 200\pi - 400 \text{ cm}^2.$$

Area – Way B

One can look at the similarity between shape 5 and shape 8. If we divide shape 5 by means of a symmetry axis which is parallel to the side of the square, then the white part of shape 8 is twice the blackened part of shape 5. Therefore, if we deduce twice the blackened part of shape 5 from the area of the square, we will obtain the required area.

$$20^2 - 2 \times (400 - 100\pi) = 400 - 800 + 200\pi = 200\pi - 400 \text{ cm}^2.$$

Perimeter – Way A

One can look at the similarity between shape 7 and shape 8. The perimeter is four times the perimeter of one "leaf". The perimeter of each "leaf" is two-quarters of the circle, namely half the circle.

In shape 8, we will divide the square into four squares so that every square circumscribes one "leaf". We will observe the reduced square which is a quarter of the large square and circumscribes a reduced shape of shape 7:

Side of the reduced square: $\frac{20}{2} = 10$ cm, the radius of the reduced circle: $\frac{20}{2} = 10$ cm.

The total perimeter of the blackened area is four times the perimeter of the reduced shape 7. Hence, everywhere the length of the square's side is 20, we will write: $4 \times \left(2 \times \left(\frac{1}{4} \times \left(2\pi \times \frac{20}{2}\right)\right)\right) = 40\pi$ cm.

Perimeter – Way B

The perimeter of the shape is the total perimeter of four half circles, namely the perimeter of two circles.

Side of the square: 20 cm, radius of the circle: $\frac{20}{2} = 10$ cm.

The total perimeter of the blackened shape is four times the perimeter of half a circle. Consequently: $4 \times \frac{1}{2} \times 2\pi \frac{20}{2} = 40\pi$ cm.

Solutions of activity 2

a) Side of the square: 20 cm, radius of the circle: $\frac{20}{2} = 10$ cm.

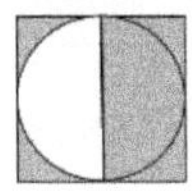

Area – the area of the square minus the area of half the sphere.

$20^2 = 400$ $\qquad$ $\frac{1}{2}\pi\left(\frac{20}{2}\right)^2 = \frac{1}{2}\pi \times 100 = 50\pi$

The total of the blackened area is the area difference between the square and half the circle. Consequently:

$$20^2 - \frac{1}{2}\pi\left(\frac{20}{2}\right)^2 = 400 - \frac{1}{2}\pi \times 100 = 400 - 50\pi \text{ cm}^2.$$

Perimeter – the perimeter of the square plus half the perimeter of the circle plus a diameter.

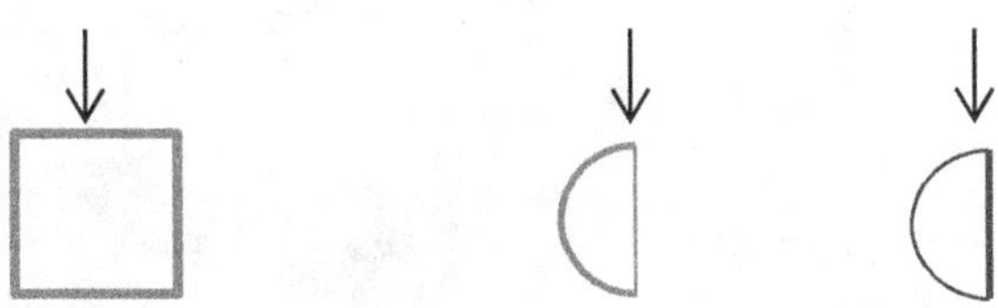

$2 \times \frac{20}{2} = 20$ $\qquad$ $\frac{1}{2}\left(2\pi \times \frac{20}{2}\right) = 10\pi$ $\quad$ $4 \times 20 = 80$

The total perimeter of the blackened shape is the amount of the perimeters of the square, half the circle, and the diameter. Consequently:

$$4 \times 20 + \frac{1}{2} \times \left(2\pi \times \frac{20}{2}\right) + 2 \times \frac{20}{2} = 80 + 10\pi + 20 = 100 + 10\pi \text{ cm}.$$

b) Side of the square: 20 cm, radius of the circle $\frac{20}{4} = 5$ cm.

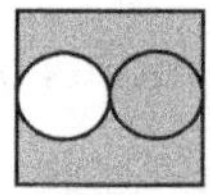

Area – the area of the square minus the area of the sphere.

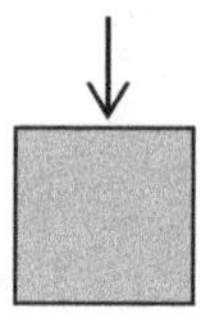

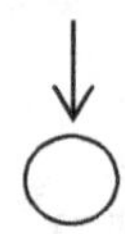

$20^2 = 400$ $\pi\left(\frac{20}{4}\right)^2 = 25\pi$

The total of the blackened area is the area difference between the square and the sphere. Hence: $20^2 - \pi\left(\frac{20}{4}\right)^2 = 400 - 25\pi$ cm².

Perimeter – the perimeter of the square plus the perimeter of the circle.

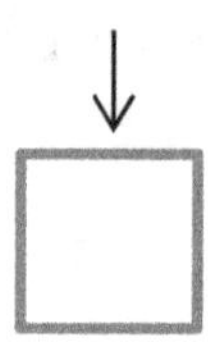

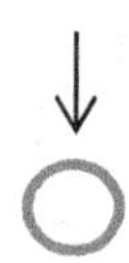

$4 \times 20 = 80$ $2\pi \times \frac{20}{4} = 10\pi$

The total perimeter of the blackened shape is the amount of the square and circle perimeter. Consequently:

$$4 \times 20 + 2\pi\frac{20}{4} = 80 + 10\pi \text{ cm}^2.$$

c) Side of the square: 20 cm, radius of the circle $\frac{20}{4} = 5$ cm.

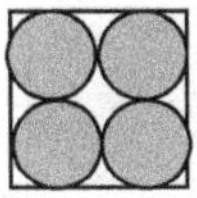

Area - Four times the area of the sphere.

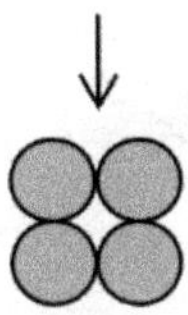

$$4\left(\pi \times \left(\frac{20}{4}\right)^2\right) = 100\pi$$

The total of the blackened area is the area of the four spheres. Hence: $4\left(\pi \times \left(\frac{20}{4}\right)^2\right) = 100\pi$ cm^2.

Perimeter – Four times the circumference of the circle.

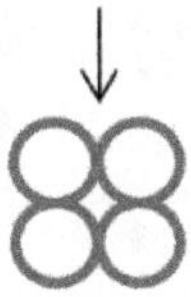

$$4\pi \times 2\pi \times \frac{20}{4} = 40\pi$$

The total circumference of the blackened shape is the total circumferences of the four circles. Consequently: $4\pi \times \frac{20}{4} = 20\pi$ cm.

d) Side of the square: 20 cm, radius of the circle $\frac{20}{2} = 10$ cm.

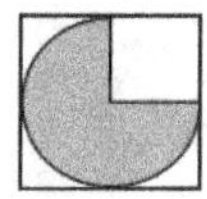

Area – three times the area of a quarter of the sphere.

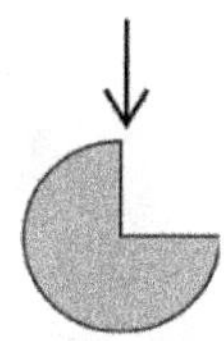

$$3 \times \frac{\pi\left(\frac{20}{2}\right)^2}{4} = 75\pi$$

The total of the blackened area is three quarters of the sphere and hence

$$3 \times \frac{1}{4} \times \pi\left(\frac{20}{2}\right)^2 = 75\pi \text{ cm}^2.$$

Perimeter – the perimeter of three-quarters of the perimeter of the circle plus two radii.

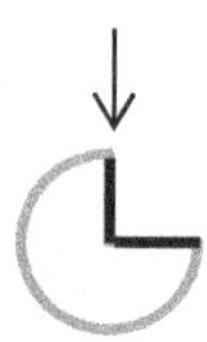

$$2 \times \frac{20}{2} = 20$$

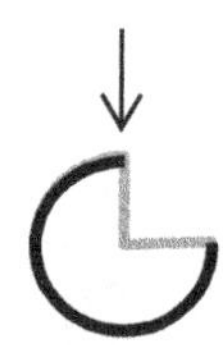

$$3 \times \frac{1}{4} \times 2\pi\frac{20}{2} = 15\pi$$

The total perimeter of the blackened shape is the number of perimeters of the three-quarters of the circle and two radii. Consequently, $\frac{3}{4} \times 2\pi\frac{20}{2} + 2 \times \frac{20}{2} = 15\pi + 20$ cm.

e) Side of the square: 20 cm, radius of the circle $\frac{20}{2} = 10$ cm.

Area – Four times one-quarter of the sphere.

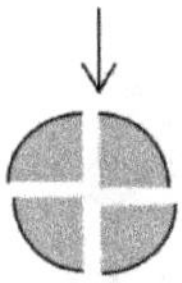

$$4 \times \frac{\pi\left(\frac{20}{2}\right)^2}{4} = 100\pi$$

The total of the blackened area is the four quarters of the sphere, which are a circle. Hence: $4 \times \frac{1}{4} \times \pi\left(\frac{20}{2}\right)^2 = 100\pi$ cm^2.

Perimeter – circumference of the square plus four quarters of the circumference of the circle.

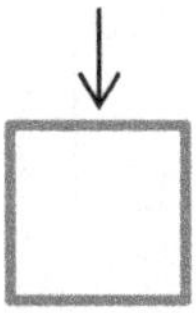

$$4 \times 20 = 80$$

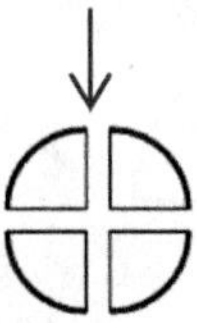

$$4 \times \frac{1}{4} \times \left(2\pi \times \frac{20}{2}\right) = 20\pi$$

The total circumference of the blackened shape is the circumference of the square and the circumference of the four quarters of the circle, namely circumference of a square and circumference of a circle. Consequently, $4 \times 20 + 4 \times \frac{1}{4} \times 2\pi\frac{20}{2} = 80 + 20\pi$.

f) Side of the square: 20 cm, Radius of the circle: 20 cm.

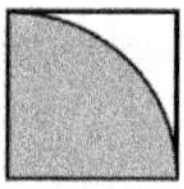

Area – Area of one-quarter of the sphere.

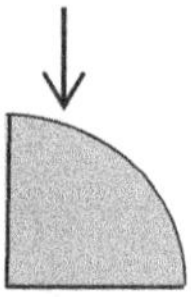

$$\frac{\pi \times 20^2}{4} = 100\pi$$

The total blackened area is one-quarter of the sphere. Hence: $\frac{\pi\times 20^2}{4} = 100\pi$ cm^2.

Perimeter – Perimeter of one-quarter of the circle plus half the perimeter of the square.

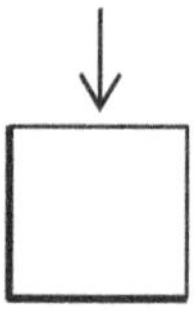

$$\frac{1}{4}\times(2\pi \times 20) = 10\pi \qquad \frac{1}{2}\times(4 \times 20) = 40$$

The total perimeter of the blackened shape is the perimeter of one-quarter of the circle plus half the square. Hence;

$\frac{1}{4} \times 2\pi \times 20 + \frac{1}{2} \times (4 \times 20) = 10\pi + 40$ cm.

g) Side of the square: 20 cm, Radius of the circle: 20 cm.

Area – area of the square minus the area of shape 7.

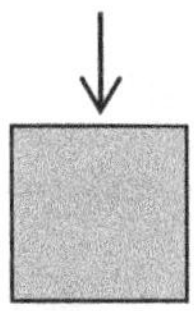

$20^2 = 40$

$200\pi - 400$

The total of the blackened area is the area difference between the square and the area of shape 7. Hence:

$20^2 - (200\pi - 400) = 400 - 200\pi + 400 = 800 - 200\pi$ cm^2.

Perimeter – Perimeter of a square plus the perimeter of two-quarters of the circle.

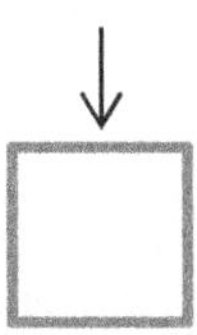

$4 \times 20 = 80$

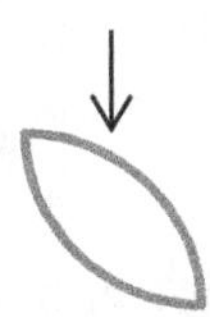

$2 \times \frac{1}{4} \times 2\pi \times 20$

The total perimeter of the blackened shape is the amount of the perimeter of the square plus the perimeter of the two-quarters of the circle, namely perimeter of a square and perimeter of half a circle. Consequently:

$4 \times 20 + 2 \times \frac{1}{4} \times 2\pi \times 20 = 80 + 20\pi$ cm.

h) Side of the square: 20 cm, radius of the circle $\frac{20}{2} = 10$ cm.

Area – Way A

The similarity between shape (h) and shape 8 is visible.

Side of the square: 20 cm, radius of the circle $\frac{20}{2} = 10$ cm.

The total of the blackened area is the area of the square minus the area of shape 8:
$20^2 - (200\pi - 400) = 400 - 200\pi + 400 = 800 - 200\pi$ cm^2.

Area – Way B

The similarity between shape () and shape (h) is visible. Shape (h) is twice shape 5:
$2 \times (400 - 100\pi) = 800 - 200\pi$.

Perimeter – the perimeter of the shape is the number of perimeters of eight quarters of circles (namely circumference of two circles) plus a perimeter of a square.

The total perimeter of the blackened shape is eight times the perimeter of a quarter of a circle plus a perimeter of a square. Hence:

$8 \times \frac{1}{4} \times 2\pi \frac{20}{2} = 40\pi$ cm.

Solutions of activity 3

The side of the square: 20 cm, the radius of the circle: $\frac{20}{6} = 3\frac{1}{3}$ cm.

Area – the area of the square minus the area of nine spheres.

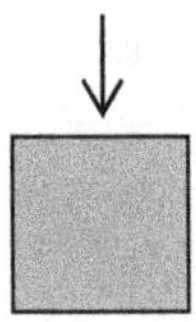

$20^2 = 40$

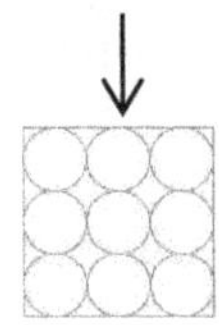

$9 \times \pi \left(\frac{20}{6}\right)^2 = 100\pi$

The total of the blackened area is the area difference between the square and the area of the nine spheres. Hence:

$20^2 - 9 \times \pi \left(\frac{20}{6}\right)^2 = 400 - 100\pi$ cm^2.

Perimeter – the perimeter of the square plus the perimeter of nine circles.

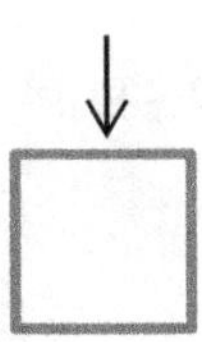

$4 \times 20 = 80$

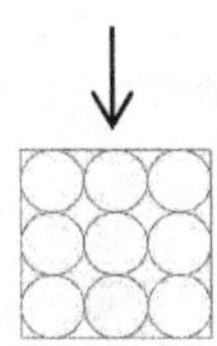

$9 \times \left(2\pi \times \frac{20}{6}\right) = 60\pi$

The total perimeter of the blackened shape is the amount of the perimeters of the square and the nine circles. Consequently:

$$4 \times 20 + 9 \times \left(2\pi \times \frac{20}{6}\right) = 80 + 60\pi \text{ cm.}$$

The side of the square: 20 cm, the radius of the circle: $\frac{20}{14} = 1\frac{3}{7}$ cm.

Area – the area of the square minus the area of 49 circles.

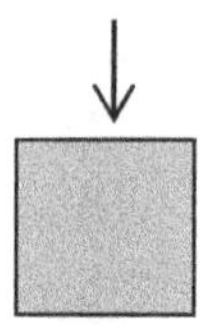

$20^2 = 40$

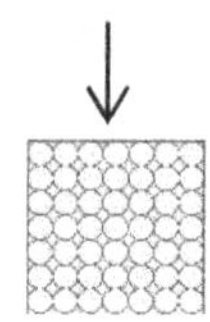

$49 \times \pi \left(\frac{20}{14}\right)^2 = 100\pi$

The total of the blackened area is the area difference between the square and the area of the 49 spheres. Hence:

$$20^2 - 49 \times \pi \left(\frac{20}{14}\right)^2 = 400 - 100\pi \text{ cm}^2.$$

Perimeter – the perimeter of the square plus the perimeter of 49 circles.

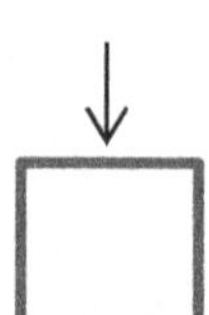

$4 \times 20 = 80$

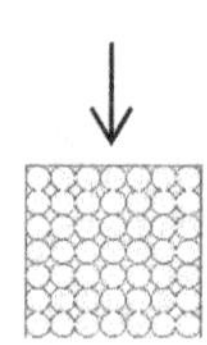

$\left(2\pi \times \frac{20}{14}\right) = 140\pi$

The total perimeter of the blackened shape is the number of perimeters of the square and the 49 circles. Consequently

$$4 \times 20 + 49 \times \left(2\pi \times \frac{20}{14}\right) = 80 + 140\pi \text{ cm.}$$

Solutions of activity 4

Side of the square: 20 cm, radius of the circle $\frac{20}{2} = 10$ cm.

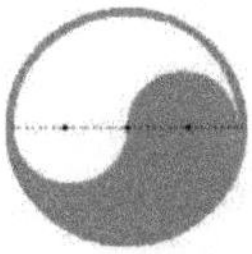

Area – half a big sphere, minus half a small sphere and another half of a small sphere, namely half a big sphere.

$$\frac{\pi\left(\frac{20}{2}\right)^2}{2} - \frac{\pi\left(\frac{10}{2}\right)^2}{2} + \frac{\pi\left(\frac{10}{2}\right)^2}{2} = \frac{\pi\left(\frac{20}{2}\right)^2}{2} = 50\pi \text{ cm}^2.$$

Perimeter – half a big sphere plus half a small sphere and another half of a small sphere, namely a total of half a big sphere and another small sphere.

$$\frac{2\pi\left(\frac{20}{2}\right)}{2} + \frac{2\pi\left(\frac{10}{2}\right)}{2} + \frac{2\pi\left(\frac{10}{2}\right)}{2} = \frac{2\pi\left(\frac{20}{2}\right)}{2} + 2 \times \frac{2\pi\left(\frac{10}{2}\right)}{2} = 10\pi + 2 \times 5\pi = 20\pi \text{ cm}.$$

Summary

All the activities presented here included individual and group experiences as well as discussions of the entire class. The verbalization prior to the graphical solution and the debates about it constituted not only a starting point for the learning but also an opportunity for enhancement and generalization at a high level of thinking. They allowed pupils at different levels, including pupils whose calculation competences were not sufficiently high, to take part in the activity. The analysis and synthesis, product analysis and the process resulted in reflective assessment and personal lessons and insights.

These activities facilitated talking with pupils about the possible relations between areas and perimeters of geometric shapes as well as leading them to draw conclusions by means of directed questions. For example:

- Does a shape with a bigger area have necessarily a bigger perimeter and vice versa?
- Can a complex shape be changed while maintaining its area or its perimeter?

Moreover, the pupils were requested to propose complex shapes obtained by the integration and synthesis of the basic shapes whose area and perimeter could be calculated according to the calculation of the areas and perimeters of the basic shapes. Of course, the additional shapes suggested by the pupils could be used as a continuation and enrichment of the activities in which they had engaged during the lessons and the discussions.

Below is an example of a word problem which relies on the basic shapes – a square and a sphere:

A shop built like a square whose side is 4 m. long was renovated. This included removing an area of half a sphere from the shop area and adding two identical complexes shaped like a hemisphere, as described in the drawing.

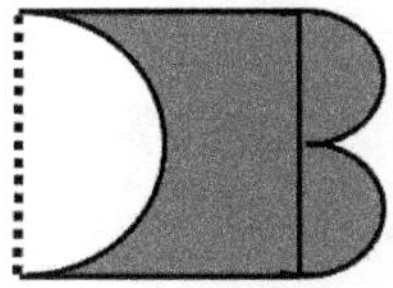

a. Has the shop area increased/decreased/not changed after the renovation?
b. Has the shop perimeter increased/decreased/not changed after the renovation?

Then the pupils can be asked to create sequences of other basic shapes.

The following are activities which integrate basic shapes. Each activity includes basic shapes of a square and a sphere. As an example of integrating additional basic shapes is presented in Activity E, whose basic shapes are a sphere and right-angled triangle. In-service teachers, pre-service teachers, and pupils who practiced the previous activities can compose a sequence of examples graded. Hence, assimilation of a creative activity of this type enables its extension and generalization.

a. The creative activity of Student A:

The radius of circle R of shape (a) is given. Calculate the area and the perimeter of the highlighted shapes.

Example (a)

Example (b)

Example (c)

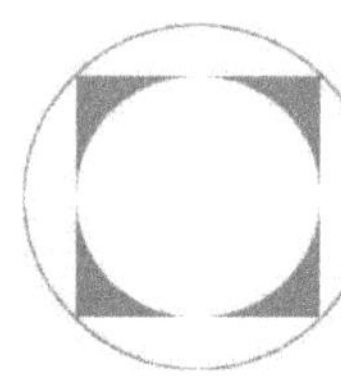

Example (d)

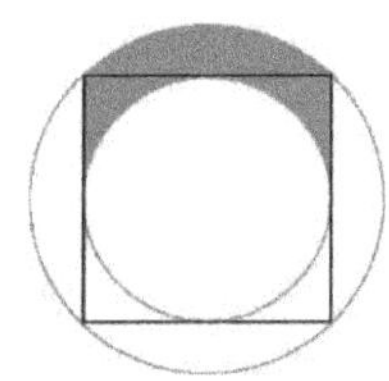

Example (e)

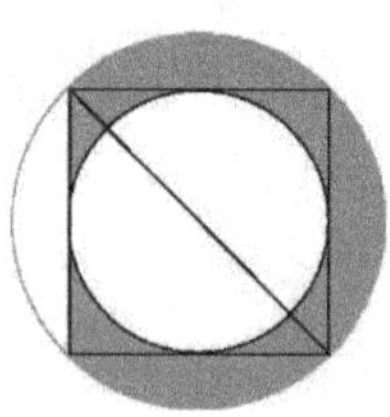

Example (f)

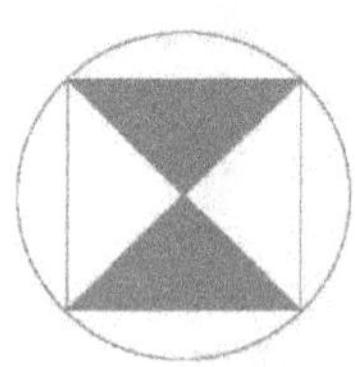

Example (g)

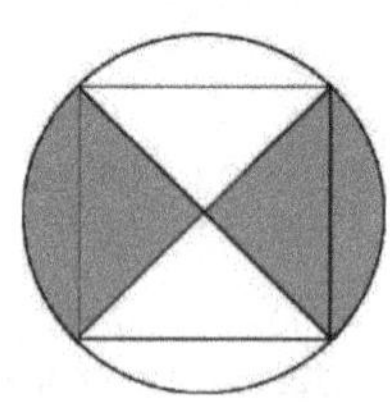

Example (h)

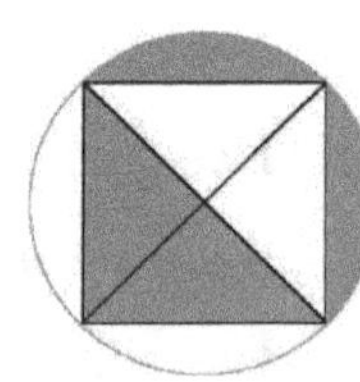

b. The creative activity of Student B:

(a)

(b)

(c)

(d)

(e)

(f)

c. The creative activity of Student C:

Example (a) **Example (b)** **Example (c)** **Example (d)**

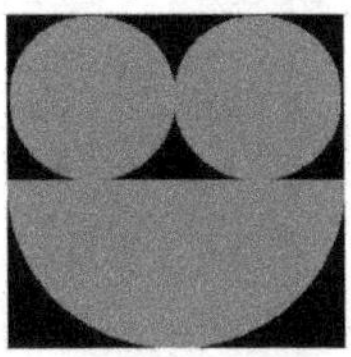

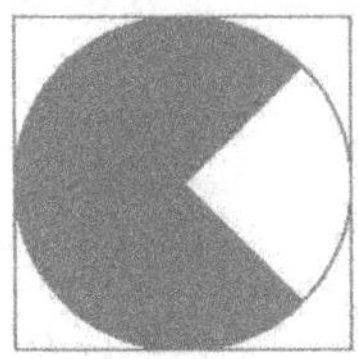

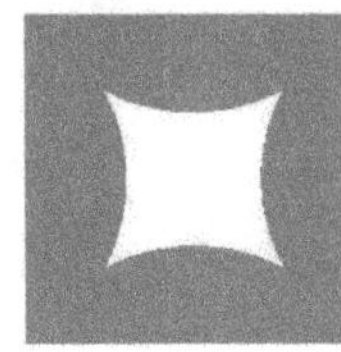

d. The creative activity of Student D:

1

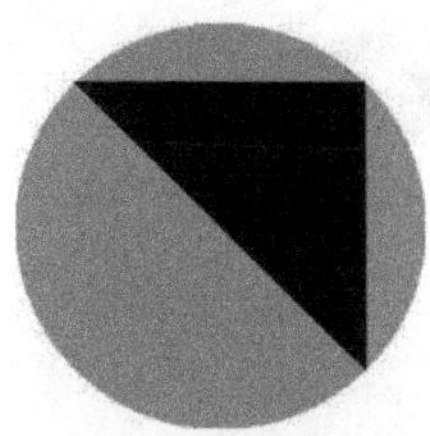

2

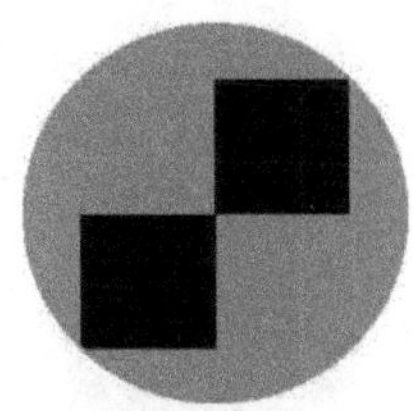

3

4

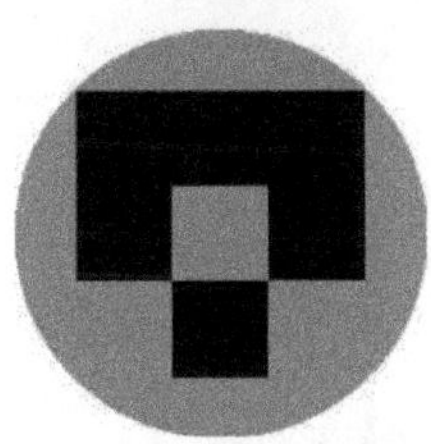

5

6

e. The creative activity of Student E:

In all the drawings, the radius of the circle is R, the short bisector in the right-angled triangle is also R and the length of the hypotenuse is 2R.

Example (a)

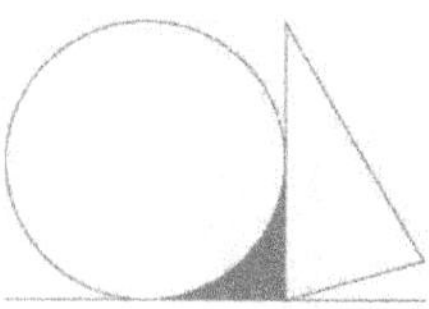

Example (b)

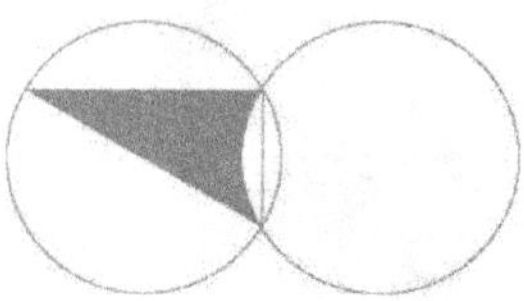

Example (c)

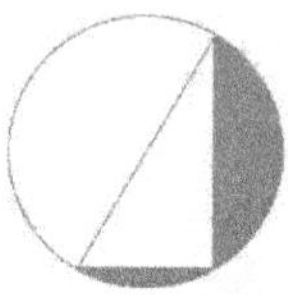

Example (d)

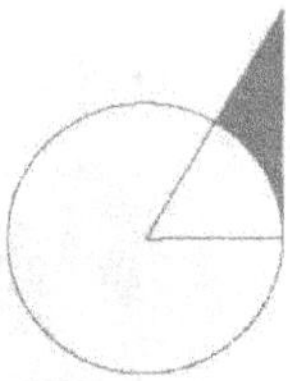

Example (e)

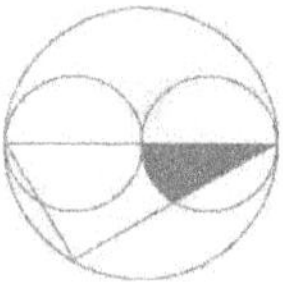

Example (f)

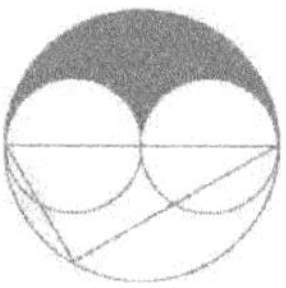

References

1. Aran, Z., & Ashkenazi, G. (2008). Creativity in raising questions as leverage for improving scientific thinking in inquiry experiments in chemistry. *Magazine about chemistry*, 13, 18–23. [Hebrew]
2. Cook, R. (2010). 'clickers' and Bloom theory in teaching. *Hed the New Ulpan for the inculcation of Hebrew and its culture*, 96, 73–38. [Hebrew]
3. De Bono, A. (1993). *Teach your child to think*. Jerusalem: Ministry of Education and Culture and Branco Weiss Institute. [Hebrew]
4. Eden, S., Mevarech, Z., & Koretz, G. (2008). The effect of meta-cognitive guidance on academic attainments in a computerised course. In: Y. Eshet-Alkalai. A. Caspi, & N. Neri (Eds.), *Book of Chase Conference for the study of learning technology: The learning person in the technological age* (pp. 115–121). Raanana: Open University. [Hebrew]
5. Grossman, S., Rogers, B., & Mor, B. (1995). *Innovation in action – Releasing creativity from its bonds*. Jerusalem: Ministry of Education and Culture and Branco Weiss Institute. [Hebrew]
6. Landau, A. (1973). *Creativity*. Tel Aviv: Cherikover Publishing. [Hebrew]
7. Ministry of Education, Culture and Sport (2006). *Mathematics curriculum for the 1st–6th grades in all sectors*. Jerusalem: Ministry of Education, Department for Curricula planning and development. [Hebrew]
8. Nevo, B. (1997). *The human intelligence*. Raanana: Open University. [Hebrew]
9. Shriki, A. (2010). Working like real mathematicians: Developing prospective teachers' awareness of mathematical creativity through generating new concepts. *Educational Studies in Mathematics,* 73, 159–179.
10. Sriraman, B. (2005). Are giftedness and creativity synonyms in mathematics? *The Journal of Secondary Gifted Education*, XVII(1), 20–36.
11. Vygotsky, L. S. (1978). *Mind in society: The development of higher psychological processes*. (M. Cole, V. John-Steiner, s. Scribner, & E. Souberman, Eds. & Trans.). Cambridge, MA: Harvard University Press.
12. Yee, F. P. (2005). Developing creativity in the Singapore primary mathematics Classes: Factors that support and inhibit. *Thinking Classroom, 6*(4), 14–46.

13. Ellerton, N. F. (1986b). Mathematics problems written by children. Research in *Mathematics Education in Australia, December,* 32–44.

14. Haylock, D. W. (1986). Mathematical creativity in schoolchildren. Journal of *Creative Behavior, 21*(1), 48–59.

15. Lavy, I., & Shriki, A. (2010). Engaging in problem-posing activities in a dynamic geometry setting and the development of prospective teachers' mathematical knowledge. *Journal of Mathematical Behavior, 29,* 11–24.

16. Levav-Waynberg, A., & Leikin, R. (2012). Using multiple solution tasks for the evaluation of students' problem-solving performance in geometry. *Canadian Journal of Science, Mathematics and Technology Education, 12*(4), 313–333.

17. Nikata, K., & Shimada, H. (2005). Facilitation of analogical transfer by posing an analogous problem for oneself. *Japanese Journal of Educational Psychology, 53,* 381–392.

18. Shriki, A. (2004). Being a real mathematician – Everyone can do it! *Alex – Bulletin for Mathematics Teachers, 31*, 4–13. [Hebrew].

19. Shriki, A. (2007). When you investigate – you discover. *Alex – Bulletin for Mathematics Teachers, 37*, 101–106. [Hebrew].

20. Shriki, A. (2013). A model for assessing the development of students' creativity in the context of problem posing. *Creative Education, 4*(7), 430–439.

21. Shriki, A., & Lavy, I. (2012). Problem posing in a dynamic geometry environment and the development of mathematical insights. *The International Journal of Learning, 18*(5), 61–70.

22. Silver, E. A. (1997). Fostering creativity through instruction rich in mathematical problem solving and problem posing. *ZDM - The International Journal on Mathematics Education, 29*(3), 75–80.

23. Silver, E. A., & Cai, J. (1996). An analysis of arithmetic problem posing by middle school students. *Journal for Research in Mathematics Education, 27,* 521–539.

24. Stoyanova, E., & Ellerton, N. F. (1996). A framework for research into students' problem posing. In: P. Clarkson (Ed.), *Technology in mathematics education* (518–525). Melbourne: Mathematics Education Research Group of Australasia.

https://doi.org/10.1142/9789811218750_0005

Chapter 5

Analytic Geometry and Logic as Promoters of Graphical Literacy

Ilana Levenberg

Introduction

Everyday events or processes can be presented in various ways, one of them being a graphical representation on a coordinate system. There are different graphical representations which enable us to get a visual picture of the information. Moreover, these representations offer an option of using them for making calculations, comparison of sizes and shifts from one representation to another.

Information presented on a coordinate system offers numerous options of mathematical activity at different thinking levels: from primary school, where pupils are first exposed to axes, and up to the higher grades of high school. Furthermore, introducing technology into mathematics studies, enables the expansion of computer-aided mathematical activities, thus facilitating graphical drawings on an axis. In the curriculum and in textbooks, axes are mainly presented with numerical values. Thus, learners are exposed to the meaning of the points indicated on the graph. Reading information from such a system constitutes an initial basis for comprehending the values of the points and the graphical descriptions.

This chapter presents activities designed to stimulate graphical literacy, creative thinking, while developing mathematical thinking, and more, thorough comprehension of the meaning of information in an axes system. This is done without any numerical values on the axes as is prevalent in most textbooks. Furthermore, being well acquainted with unusual graphical representations will greatly enhance the study of algebra, analytic geometry, and other mathematics subjects later on.

The presented activities were given to learners from the 5th and 6th grades of primary school, the 7th grade of junior high school and to a group of pre-service teachers, specializing in mathematics teaching as part of a didactic workshop. The experience was new to all of them. In fact, their mastery of reading 'non-numerical' information was good, yet they found it difficult to decide whether the assertions were correct. Moreover, the learners encountered difficulties while presenting data on new axes, as required in activities of the highest level. As mentioned, due to the fact that the topic was new to all the participants, it evoked interest and inquisitiveness more than a regular activity of reading data from a graphical representation. The pre-service teachers indicated mainly the importance of such an experience and its contribution to an in-depth understanding of information reading, namely 'graphical literacy'.

Theoretical Background

Visual literacy and creative thinking development in mathematics lessons

The topic of visual literacy is an inseparable part of geometry, algebra and analytic geometry studies. Literacy, defined as reading, writing and speaking. encompasses in fact also a visual ability. This implies the ability to use visual representations (images), attribute a meaning to them, as well as learn to express what one sees in the image by applying the appropriate language and the suitable terms. This is a cognitive ability that can be developed, like reading or writing. Due to the rapid digital development in the 21st century, literacy in general and visual literacy in particular, have become the focus of the teaching theory and, hence, its importance to the learners is unquestionable.

According to Kedra and Zakeviciute (2019), visual representations must be presented by a "lingua franca" in daily communication. At present, one cannot describe learning and acquisition of any knowledge, including mathematics knowledge, without visual representations. The transition from verbal to visual messages, greatly facilitates understanding of the learned material, in parallel to the development of ideas and creative thinking.

Visual literacy is not acquired in an immediate or independent manner. It depends on the learning subject, its language and the terms associated with it. Highlighting the development of visual literacy is essential particularly in a subject like mathematics, considered as abstract and hard, with its own unique language. The development of learners' visual literacy will serve them during their studies of any subject of mathematics. For them, it will constitute an instrument of mediation between the verbal and the visual and vice versa, enhancing comprehension of the studied topic. At a later stage, visual literacy will allow transitions between various visual representations. The visual literacy skills entail visual thinking and ability to master the knowledge of the various representations, all these together combined with the sense of sight (Kedra, 2018; Serafini, 2014). The development of visual ability is, in fact, a preliminary and basic step towards the development of creative thinking, which all teaching practitioners strive to attain.

Research of mathematics education (Gazit & Patkin, 2009; Leikin, Berman, & Koichu, 2009; Shriki, 2010; Levav-Waynberg & Leikin, 2012), reinforces the need for seeking various ways of developing mathematical thinking and creativity. One of the ways of promoting this issue is the presentation of exceptional activities that require 'out of the box' mathematical thinking (Levenberg, 1995, 2015).

Teaching mathematics at primary school usually focuses on the inculcation of algorithms and, through them, getting the pupils acquainted with logical thinking which complies with the rules of mathematics. All these leave mathematics teachers a little time and latitude for open-ended activities that facilitate the expression of creative thinking. In fact, only when teachers' objective is to develop higher-order thinking, they should implement some of the recommendations made by Resnick (1987). For example: design assignments which allow learners to find several solutions, activities requiring the application of different criteria or activities which are not algorithm-based.

In his paper, Mann (2006) reviews the different definitions of creativity in mathematics. He argues that when learners graduate, they are adequately capable of calculating and performing algorithms but are unable to apply this knowledge in a meaningful way. According to him, teaching mathematics with no investment in the promotion of creativity does not allow learners to see the beauty of mathematics. Moreover, it prevents talented and gifted learners from developing their special skills in this subject.

To sum up, the frameworks of mathematical activities prevalent in class, as well as their levels, are dictated by the textbooks. The way they are presented to learners depends mainly on mathematics teachers. The higher teachers' awareness of developing mathematical and creative thinking is, the better they know how to cope with the challenge.

Graphical representations as part of visual literacy and the teaching thereof

Since the beginning of the 21st century, graphical representations have become much more common and available in any field, not only in mathematics and other sciences. New technological instruments allow an almost immediate translation of information about varied topics into different graphical representations. Shah & Freedman (2011) showed that between the years 1984–1994, the number of graphical representations in academic magazines has doubled. This is demonstrated also in daily newspapers. Graphs have turned into the tangible, available and easiest 'instrument'for presenting information of all types. Consequently, it has become necessary to enhance the issue of teaching graphs at schools and higher education institutions, as part of the inculcation of scientific literacy.

Moreover, scientific research of graphs has increasingly expanded. For example, Berg & Smith (1994) investigated the way learners read and explain information presented by a graph. They identified a significant relation between learners' logical thinking capability and graphing skills. Other studies (Canham & Hegarty, 2010; Cook, Wiebe & Carter, 2008; Hipkins, 2011) explored what affects learners' mastery of graphs. The findings illustrate that, in addition to the learners' thinking capability, educators' teaching strategies and learning to use computers while teaching the subject, are also important. All these factors combined together have an impact on the learners' capability of creating new graphs and explaining the information presented to them.

The study conducted by Carpenter & Shah (1998) presents the three stages of becoming versed in the issue of graphs. First, learners should be acquainted with the visual part, namely the representative shape of the graph, the importance of the shape and its representative features. Second, they have to know the numerical and quantitative components and what they represent. Third, they should connect the quantitative and graphical parts of the conclusions.

As to the determination of learners' knowledge of graphs, Wainer (1992) suggests the type of questions which can be asked: (a) Basic level questions which relate to data extraction; (b) Intermediate level questions which engage in trends emerging from the information; and (c) high-level questions which deal with the depth of the information meaning while comparing trends.

To sum up, the research literature (Friel, 2001; Shah & Freedman, 2011; Wang, 2012) classifies the information presented by graphs into three types which also represent the various thinking levels required by learners, as follows:

1. Explicit information – this information is usually presented already in the axis names and it is easily identified.
2. Tacit information – for obtaining this information, learners should examine more thoroughly what is represented by the graph. For that purpose, mathematics knowledge (basic or advanced) is necessary.
3. Conclusive information – the highest level of information which requires reasoning and justification by means of mathematical instruments in order to draw the conclusions.

Acquaintance with René Descartes as an Introduction to the Teaching of the Cartesian Coordinates

"Cogito ergo sum" – *I think, therefore I am* (René Descartes).

René Descartes, the French mathematician and philosopher (1596–1650), was the first to come up with the idea of accurately presenting the location of points on the plane and, later, in space. He conceived the idea while serving in the army as an officer on the banks of the Danube. Some say that the innovation was the result of a dream-vision. Descartes delayed publishing his invention and only in 1637 he printed his treatise *Discourse on the method* – a method through which he presented his philosophical concept "The method of leading intelligence in the right way". At the end of the treatise, there are three appendices. The third appendix which is 106-page long is entitled: "Geometry". It includes Descartes' greatest contribution to science: every point on a plane can be presented by means of a pair of real numbers. The mathematician and philosopher Leibnitz invented calculus at the same time as Newton during the second half of the 17th century and called it on the name of the person who conceived it: "Cartesian Coordinates System" (Gazit, 2004).

The "Cartesian Coordinates System" – is derived from Descartes, a name which is pronounced 'Cartesius' in Latin. The Cartesian Coordinates System is defined by two straight lines called axes which are usually positioned at a right angle. The horizontal axis is indicated by X and the vertical axis by Y. The point of intersection of the axes is called the *origin of the axes*. In order to indicate a certain point on the axes system, we indicate the values of the X and the values of the Y on the point, thus creating the ordered pair of numbers (X,Y). The expression "Cartesian Coordinates System" serves only in 2- and a 3-dimensional space.

This system paved the way to analytic geometry and modern mathematics. It is inconceivable to imagine the existence of numerous developments in mathematics and their implementation without the

Cartesian Coordinates System. Hence, we should start teaching this topic by first presenting Descartes and his extensive contribution to mathematics in general and to the topic of axes in particular. The narrative behind the invention of the Cartesian Coordinates System and their mathematical development in future generations should be the basis for the study of this essential topic. Teachers can tell learners about Descartes who saw a fly on the wall and conceived the idea of an axes system as a mathematical instrument. By means of this instrument, one can present the walking course of the fly and the possibility of presenting its location on the wall at any given moment.

Examples of Activities Based on Reading Information from a Graph

The following activities promote and enhance mastery of reading information presented by axes. The activities bridge between mathematics learned in class and occurrences in the learners' environment. Moreover, this kind of activities can facilitate a better understanding of the topic of functions which is going to be studied later.

For the purpose of presenting information or reading it from a graph, learners need first to be acquainted with axes and understand the meaning of the graph and its various representations. In order to draw conclusions based on a graphical representation, mathematics knowledge, as well as logical thinking capability, are also required.

Preliminary activity

Please write in words what is presented in the graph below, your conclusions and insights.

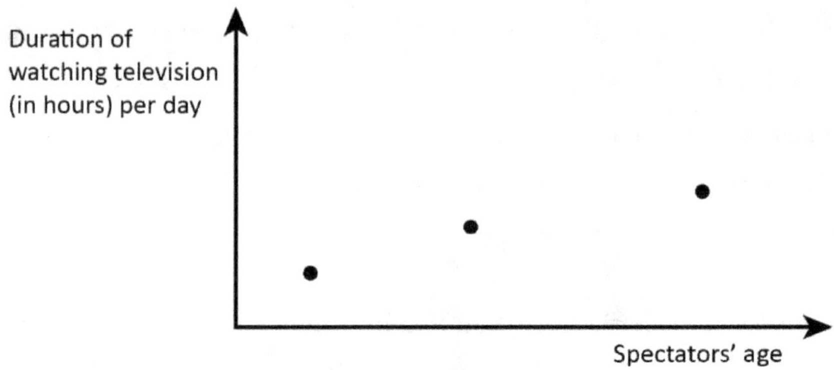

Acquaintance with and comprehension of the above graph, while writing or verbalizing things, enables moving to the next activities which are more complex and require coping with higher-level assignments.

Suggestions for teaching the subject in class

a. Ask the learners to 'tell', describe in writing or orally what is the narrative of the graph.
b. Ask the learners to 'tell', describe in writing or orally each of the assertions and indicate which of them are correct and according to which graphs.
c. Ask the learners to raise additional questions for a given graph.
d. Present examples from the press or from everyday life, in order to describe a real graph.

Activity No. 1 is the easiest and requires only initial and basic knowledge of reading information. Activity No. 2 requires higher-level thinking as well as a capability of formulating assertions and drawing conclusions about a graph that is not explicit. Here too the issue of creative thinking development is integrated.

Activity No. 1

Below are graphs that describe information about two snacks, 'Knight' and 'Champions'.

The '**Knight**' snack bag is marked with the letter **K** and the '**Champions**' snack bag is marked with the letter **C**.

Graph No. 1:

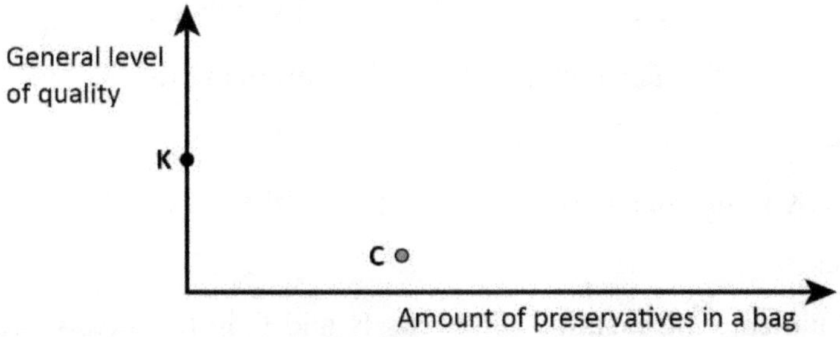

Graph No. 2:

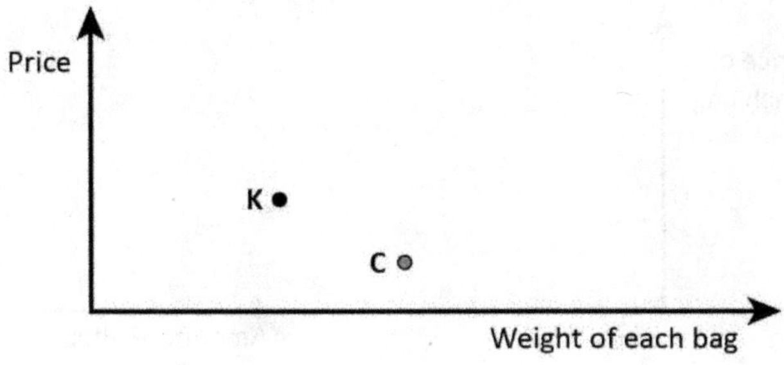

Graph No. 3:

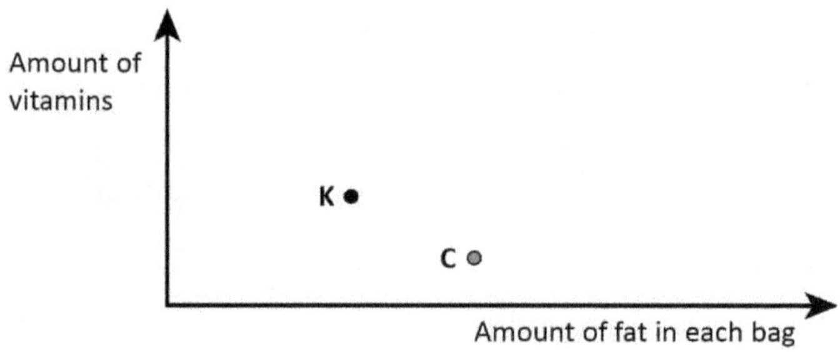

I. Check each of the following assertions and indicate which of them are correct and according to which graphs?
   1. The 'Knight' snack is cheaper than the 'Champions' snack.
   2. The 'Knight' snack bag weighs more than the 'Champions' snack bag.
   3. The 'Knight' snacks are better for weight watchers.
   4. The 'Knight' snack contains more vitamins than the 'Champions' snack.
   5. The 'Knight' snack is of higher quality than the 'Champions' snack.

II. Please indicate the location of snacks **K** and **C** in the axes system, according to the information obtained from the previous graphs.

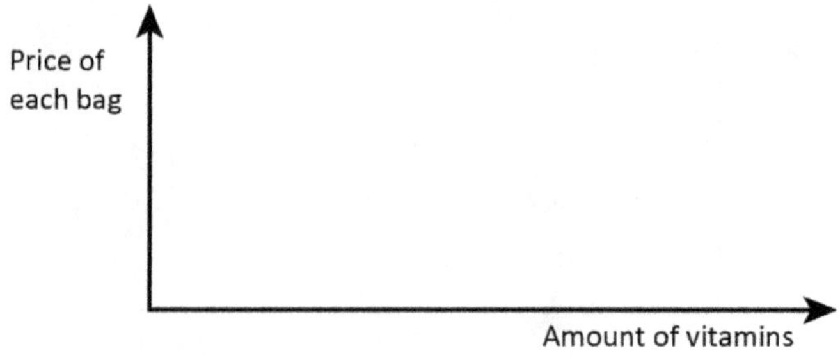

Activity No. 2:

The following graphs describe facts associated with two shopping centers, 'Knights' Mall and 'Princes' Mall. The '**Knights**' Mall is marked with the letter **K** and the '**Princes**' Mall is marked with the letter **P**.

Graph No. 1

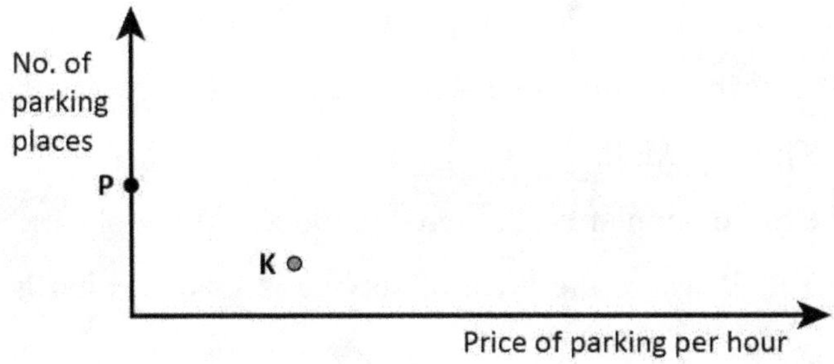

Graph No. 2

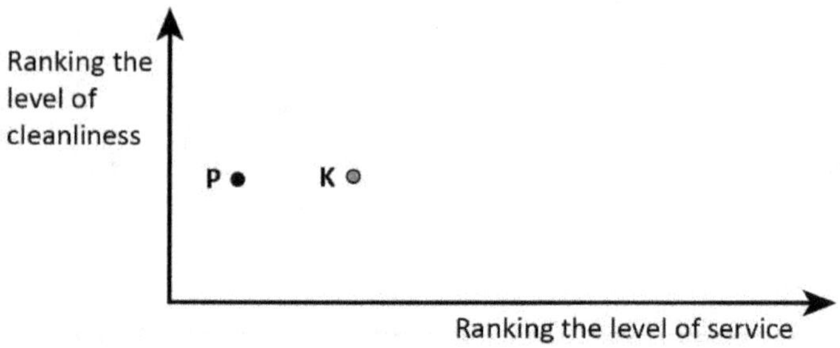

Graph No. 3

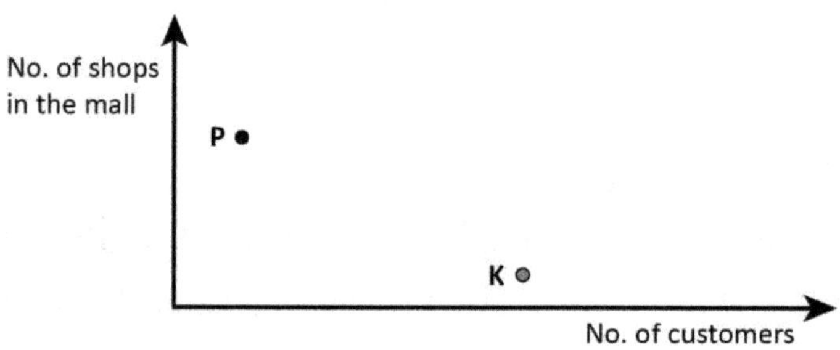

I. Check each of the following assertions and indicate which of them are correct and according to which graphs?

1. In the 'Princes' Mall the level of cleanliness is lower than that in the 'Knights' Mall.
2. In the more popular mall, there are more parking places.
3. In the mall where the level of service is low, parking is free of charge.
4. In the mall with a small number of parking places, the number of customers is higher.

II. Here is another fact: in the mall where the number of shops is greater, the prices are higher.

III. Use the information in the previous item and in the previous graphs in order to indicate the malls (points P and K) in the following axes systems.

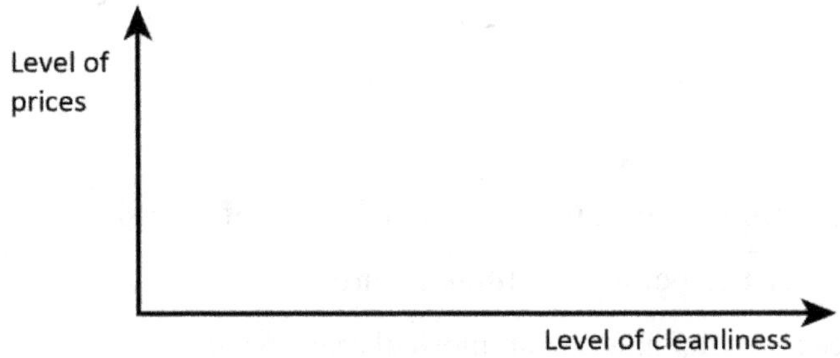

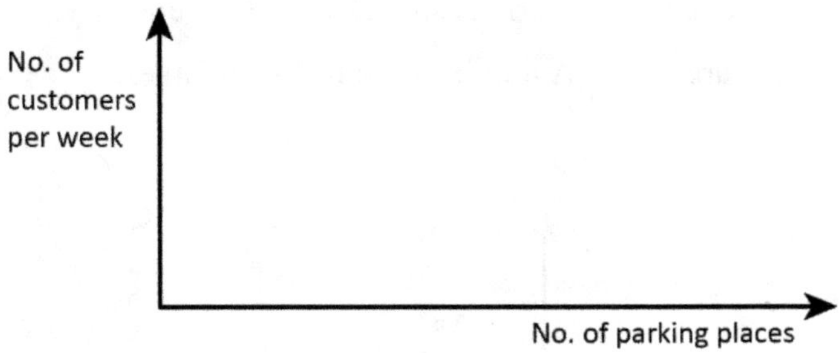

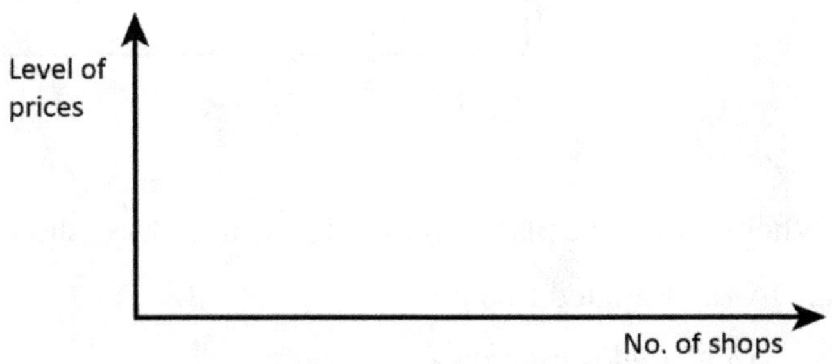

a. Based on all the information you have about the two malls, please formulate three assertions which are correct and three assertions that are incorrect.

b. Which assertions were easier to formulate, the correct or the incorrect ones?

Please give reasons for your answer.

Integration of the Foundations of Geometry and Analytic Geometry

The activities below constitute an initial basis for understanding graphical representations in a non-numerical axes system.

Now we can use the same way of inculcating graphical representation for the presentation of various shapes of plane geometry.

Below are points A and B in the following axes:

1. Where should we place a third point C in order to draw triangle ABC?
   a. A right-angled triangle.
   b. An isosceles triangle
   c. Obtuse-angled triangle

2. Where should we sketch the third point C so that no triangle can be drawn? How many options are there?
3. Three points-on an axes system are given, A, B, C.

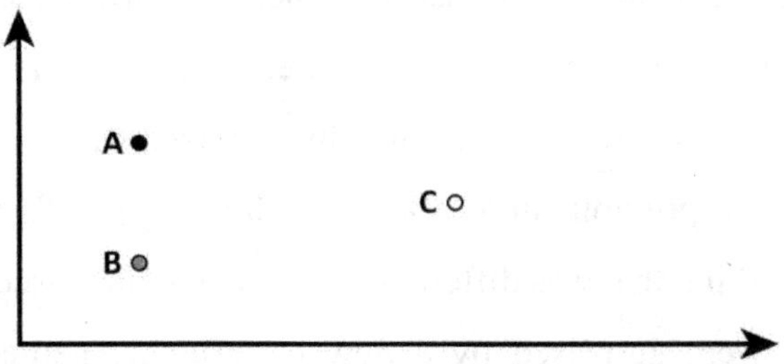

1. Where should we sketch a fourth point D in order to obtain a convex quadrilateral? Please provide an explanation and arguments.
2. Where should we sketch a fourth point D in order to obtain a concave quadrilateral? Please provide an explanation and arguments.
3. How should we sketch points A, B, C, so that they form three vertices of a square? Of a rectangle? Of a right-angled trapezoid?

These activities constitute another way of looking at necessary or optional conditions for representing various shapes of plane geometry. In fact, they enhance the properties that these shapes should have according to their definition. It is noteworthy that this chapter presents only the use of the first quadrant for the initial acquisition of orientation of an axes system. Later on, these geometrical activities can be extended to the other quadrants, with everything it implies.

Summary

The activities presented in this chapter aim to enhance the promotion of logical and creative thinking and the development of visual literacy. As indicated above, through these activities, the 5th-graders, 6th-graders, 7th-graders and the pre-service teachers learned the meaning of data in an axes system by means of a different approach. The difficulty of creating new graphs along these lines by themselves, stemmed mainly from the learners' previous inexperience with this type of activity, requiring a way of thinking that was different from the one they used in the past. Moreover, they were surprised by a new experience, namely the presentation of two-dimensional shapes on an axis. This was due to the fact that they had not yet learned analytic geometry in an orderly manner.

In the following assignment that was given to all the participants, they were requested to design a similar activity, taken from another area/topic. Most of them repeated the structure of activities that they had experienced before with only minor changes. Nevertheless, each of the research populations comprised a small number of learners who presented activities that were more difficult and complicated than the ones given to them. They expanded and integrated additional knowledge that was necessary for building new graphs, e.g. presenting three 'shopping malls' instead of two, like in the above example. This implied formulating many and more complex assertions.

Learning the topic of graphs and their meaning not by calculating numerical values, stimulated the need for a different thinking and different approach to the familiar contents. Furthermore, the learners had to formulate correct and incorrect assertions and to acquaint themselves with 'graphical literacy' that they had not experienced before. In fact, this implements the words of Silver (1997), who argued that posing problems and solving them is the heart and core of mathematical practice and creativity. The transition from a preliminary and easy activity of reading information from a graph to more complex activities, whereby learners are required to build activities in this spirit by themselves, integrate the two aspects: developing mathematical thinking and creativity.

References

1. Berg, C.A., & Smith, P. (1994). Assessing students' abilities to construct and interpret line graphs: Disparities between multiple-choice and free-response instruments. *Science Education, 78(6),* 527–554.
2. Canham, M., & Hegarty, M. (2010). Effects of knowledge and display design on comprehension of complex graphicalals. *Learning and Instruction, 20,* 155–166.
3. Carpenter, P.A., & Shah, P. (1998). A model of the perceptual and conceptual processes in graph comprehension. *Journal of Experimental Psychology: Applied, 4*(2), 75–100.
4. Cook, M., Wiebe, E.N., & Carter, G. (2008). The influence of prior knowledge on viewing and interpreting graphicalals with macroscopic and molecular representations. *Science Education, 92*(5), 848–867.
5. Friel, S. N. (2001). Making sense of graphs: Critical factors influencing comprehension and instructional implications. *Journal for Research in Mathematics Education, 32*(2), 124–158.
6. Gazit, A. (2004). *Eureka...! About people who loved to think and calculate*, Kfar-Sava: Geist Publications. [Hebrew]
7. Gazit, A., & Patkin, D. (2009). The place of creativity in the solution of exceptional problems of series among elementary school mathematics teachers and pre-service teachers in other disciplines. *Mispar Hazak* (2000), *17*, 16–24, Haifa: University of Haifa. [Hebrew]
8. Hipkins, R. (2011). Challenges with graph interpretation: A review of the literature. *Studies in Science Education, 47*(2), 183–210.

9. Kedra, J. (2018). What does it mean to be visually literate? Examination of visual literacy definitions in a context of higher education. *Journal of Visual Literacy, 37*, 67–84. doi:10.1080/ 1051144X. 2018. 1492234
10. Kedra, J., & Zakeviciute, R (2019). Visual literacy practices in higher education: what, why and how?, *Journal of Visual Literacy,* DOI: 10.1080/ 1051144 X.2019.1580438. Retrieved online: https://doi. org/ 10.1080/ 1051144X.2019.1580438
11. Leikin, R., Berman, A., & Koichu, B. (Eds.) (2009). *Creativity in mathematics and the education of gifted students*. Rotterdam, the Netherlands: Sense Publisher.
12. Levenberg, I. (1995). *Graphs. Workshop activity within the framework of the project for the promotion of science and technology studies in the Galil.* Haifa: Israel Institute of Technology, Department of Technology and Sciences Teaching. [Hebrew] Levenberg, I. (2015). Information on Coordinate System as a tool for developing mathematical thinking. *International Journal of Contemporary Educational Research (IJCER). 2(1), 46–53.* Retrieved online: http://ijcer.net/article/ view/ 5000179385
13. Levav-Waynberg, A., & Leikin, R. (2012). *The role of multiple solution tasks in developing knowledge and creativity in geometry. Journal of Mathematical Behavior, 31*, 73–90.
14. Mann, E. L. (2006). Creativity: The essence of mathematics. *Journal for the education of the gifted, 30*(2), 236–260.
15. Resnick, L. (1987). *Education and learning to think.* Washington DC: National Academy Press.
16. Serafini, F. (2014). *Reading the visual: An introduction to teaching multimodal literacy*. New York, NY: Teachers College Press.
17. Shah, P., & Freedman, E. G. (2011). Bar and line graph comprehension: An interaction of top-down and bottom-up processes. *Topics in Cognitive Science, 3*(3), 560–578.
18. Shriki, A. (2010). Working like real mathematicians: developing prospective teachers' awareness of mathematical creativity through generating new concepts. *Educational Studies in Mathematics, 73*(2), 159–179.

19. Silver, E. A. (1997). Fostering creativity through instruction rich in mathematical problem solving and problem posing. *ZDM - The International Journal on Mathematics Education,* 29*(3).75-80.*Wainer, H. (1992). Understanding graphs and tables. *Educational Researcher, 21*(1), 14–23.
20. Wang, Z. H. (2012). Students' Cognitive Reasoning of Graphs: Characteristics and progression. *International Journal of Science Education, 34(13), 2015–2041.*

https://doi.org/10.1142/9789811218750_0006

Chapter 6

Building Geometric Understanding – Identifying and Refuting False Claims

Dorit Patkin

Introduction

Proving and refuting claims are important for building geometric understanding. Imre Lakatos (1976), philosopher of mathematics in the 20th century, argues that the refutation of mathematical claims enhances the definitions of the terms on which claims are based and also the accuracy of the claim formulation. Based on her teaching experience, the writer of this chapter has realized that when teaching geometry courses, mathematics high school teachers and college and university lecturers, should pay more attention to the identification of false claims and the reasoning of their refutation. Moreover, they should be aware of the fact that there are some factors that affect the type of claims as Yopp (2015) found in his study. Researchers (see Fischbein & Kedem 1982; Wohlgemuth, 2003) maintain that identification of true claims and their proof as well as identification of false claims and their refutation are an essential instrument in building mathematics knowledge. Balacheff (1991) portrayed six ways learners use for refuting false claims. In his research, he found that the learners' thought was insufficient for using counterexamples in order to refute false claims. Hanna (2000), too, stipulates that coping with the identification of claims and attempting to prove or refute them are important not only to mathematicians but also to pre-service and in-service teachers. Potari, Zachariadea, and Zaslavsky (2010) specify that evaluating and refuting learners' claims are instrumental in teachers' practice. This implies that teachers are frequently required to give improvised suitable explanations, using examples and counterexamples.

Furthermore, proofs and refutations of claims are attributed to a key place in the *Principles and Standards for School Mathematics* (NCTM 2000). The NCTM document indicates that the concept of proof must be part of the learners' accumulated experience, from pre-school until their high school graduation. "Reasoning and proof should be a consistent part of students' mathematical experience in pre-kindergarten through grade 12" (NCTM 2000, p. 56) and "…it is noted: Students should become increasingly able to use deductive reasoning to establish or refute conjectures and should be able to use established knowledge to deduce information about other situations" (p. 310). Moreover, this document points out that abilities associated with proof or refutation should be cultivated, in order to provide learners with instruments for explaining why the claims are true or false. In view of the above, this chapter explores the extent to which pre-service teachers can identify and refute false claims.

Coping with the identification of true and false claims constitutes also an important instrument for the promotion of creative thinking. Facing states of cognitive conflict stimulates and enhances motivation (Johnson & Johnson, 2009). Learners' dealing with states of conflict, required for presenting their reasoning, should be versed in the concerned topic and be acquainted with terms and theorem for the justification. All these together call for an intellectual effort on the part of the learners. Moreover, it is necessary to mobilize that cognitive 'spark' that teachers should encourage, every time in a new way, through a challenging activity.

Inductive Reasoning, Deductive Reasoning, and Mathematical Proofs

Inductive reasoning is a way of thinking, according to which a certain claim is true on the basis of particular cases. Deductive reasoning is an inverse reasoning process, whereby the truth of the claim is determined by a set of assumptions and inferences that are derived from them and are arranged in a logical construct of proof, devoid of internal contradictions (Patkin, 2012). That is, deductive proof is a logical sequence of claims leading to a conclusion, being based only on terms and fundamental assumptions (which do not have to be proven) as well as conclusions that were drawn previously on the basis of the terms and fundamental assumptions. "Deduction is the process of deriving the consequences of what is assumed. Given the truth of the assumptions, a valid deduction guarantees the truth of the conclusion, a valid argument in which the conclusion follows the premises" (Patkin. 2012, p. 986).

Studies illustrate that in order to be able to build a deductive proof of the claim for proving its truth, learners should initially undergo a thinking process. For the purpose of describing this thinking process, Eco (1983) and Peirce (1960) conceived models based on a preliminary stage of the thinking process which presented assumptions and ideas. As Eco said:

"The real problem is how to figure out both the Rule and Case at the same time, since they are inversely related, tied together" (p. 203).

Wohlgemuth (2003) argues that the ability to prove and evaluate a proof of a true claim, is the last stage in the hierarchy of proofs, while the inductive reasoning is presented at earlier stages. The researcher stipulates that it is impossible to reach advanced stages without passing and 'maturing' (even for a long time) in the initial stages of the hierarchy. This is manifested by the fact that some learners can make assumptions based on observation but for years are unable to logically validate their assumptions.

Proof and Refutation of Claims

The empirical literature extensively discusses the difficulties learners encounter in the perception and comprehension of the essence of a claim (Hanna 2000; Vinner, 1997; Yackel & Hanna, 2003; Yollesh & Dreyfus, 2003). Studies that engaged in methods of proof used by post-primary school learners indicated that the latter had difficulties to prove claims in geometry (Clements & Battista, 1995; Herbst, 2002). Few learners were aware of the functions of the proof and of valid arguments and explanations about true claims (Fischbein & Kedem, 1982; Furinghetti, Olivero, & Paola, 2001; Vinner 1983; Waring 2000). These difficulties include also lack or partial understanding of the conceptions of counterexamples among primary school mathematics in-service teachers (Barkai, Tsamir, Tirosh, & Dreyfus, 2002), high school mathematics in-service teachers (Barkai, 2009), and pre-service teachers (Zeybek 2016). Mathematical claims identified as false can be refuted by counterexamples or by reliance on theorems known to the pre-service teachers. Counterexamples can be drawings or a rule which describes those counterexamples. For instance, in order to refute the claim "every triangle is an acute-angled triangle", it is sufficient to present a counterexample of an obtuse-angled triangle or a right-angled triangle.

Zazkis and Chernoff (2008) introduced the notions of pivotal examples and bridging examples and highlighted their role in creating and resolving a cognitive conflict. They discussed the issue of what makes a counterexample exemplary. The researchers maintained that "the convincing power of counterexamples depends on the extent to which they are in accord with individuals' example spaces" (p. 195). Fischbein (1982) and Chazan (1993) investigated the way learners prove or refute claims of the 'every' and 'there exists' type. They found that in both cases, learners succeeded to prove or refute claims by providing counterexamples. According to Fishbein (1982), most learners attest that checking particular cases rather than using formal reasoning convinces them that the claim is true. In many circumstances, after succeeding to understand a formal proof, the learners had also to check particular cases in order to be persuaded that the claim was true. These findings are in line with Patkin (2012). The key question in her study was whether 10th graders who were in the advanced mathematics pathway and studied geometry as a deductive subject knew when an example was sufficient for proving or refuting geometrical claims. Moreover, Patkin (2012) explored whether some of these learners tended to refute any claim by deductive proof even if one counterexample was sufficient. This feature was characterized and referred to by Messinger (2006) as 'over deductive'. She indicated that unlike learners she defined as ‘deductive pupils’ – namely pupils who knew when a proof was necessary and when a counterexample/example was sufficient in order to refute or justify a claim – pupils demonstrating ‘over deductive’ were unable to make this distinction.

Zaslavsky and Ron (1998) investigated in which way learners used a counterexample for refuting a mathematical claim and what were the difficulties they encountered when refuting claims by means of a counterexample. According to their findings, the learners did not comprehend that one counterexample was sufficient for refuting the claim. For example, in order to refute the false claim: “Every two right-angled triangles are similar”, it is sufficient to cite a specific counterexample, e.g. a drawing or verbal description of a right-angled triangle with an acute 20° angle and a second right-angled triangle with another 26° angle. Moreover, this claim can be refuted by the general example, namely equality of one angle is insufficient for two triangles to be similar. The equality of another angle is required for complying with the theorem of similar triangles – “angle, angle” (A A).

In a wide-scope study, Lin (2005) describes teaching and learning strategies for better instructing the pupils how to prove and refute mathematical claims. Lin classified the refutations into different categories and found that in the case of mathematical claims, quite a large number of learners could give correct estimates in the processes of proof/refutation assignments. This attests to a continuity between a refuting process and the presentation of true claims by some learners, and this continuity can be expanded to their processual proofs. Certain learners have given counterexamples together with analytic or mathematical proofs, for the purpose of refuting claims. Similar difficulties of understanding the issue of proofs and the shift to formal proofs were found in studies of higher education learners majoring in mathematics (Dreyfus, 1999; Jones, 2000; Weber, 2002, 2008). Furthermore, Potari et al. (2010) explored the way high school mathematics teachers explained to the learners how to refute false claims. Their findings identified two key approaches to the refutation suggested by these teachers: a. using a theorem with which the learners were acquainted; b. using counterexamples. The difficulties encountered by these teachers were evident in the generation of correct counterexamples as well as in the narrow meaning of theorems and the use thereof for refuting false claims. In addition, Potari et al. (2010), displayed a tree diagram with two main branches and sub-branches of reasoning, divided into categories and sub-categories.

Concerning the activity with familiar and unfamiliar false claims, the theoretical background illustrates that, despite the numerous studies of proving true claims or refuting false claims, there are still many open questions. These questions are related to the characteristics of the false claims. Consequently, mathematics educators do not have a practical approach to integrating a systematic training program, dealing with claims as part of the mathematics curriculum. This chapter explored the success in refuting false claims of different types and sub-types. The assumption was that the identification of subtypes of false claims might facilitate designing a dedicated training program for teaching students. The false claims are divided into two types: "familiar" – type A and "unfamiliar" –

type B. The term "familiar" refers to claims that the pre-service teachers have previously studied in their geometry lessons, while the term "unfamiliar" refers to claims that they have not studied before their participation in this study. The "familiar" false claims (type A) can be divided also into three sub-types: An incorrect generalization of a familiar true claim, i.e. a theorem (type A_1); a faulty and inaccurate formulation of a familiar true claim (type A_2); or an incorrect opposite claim of a familiar true claim (i.e. a theorem) (type A_3). Figure No. 1 summarizes the types of false claims discussed in this chapter.

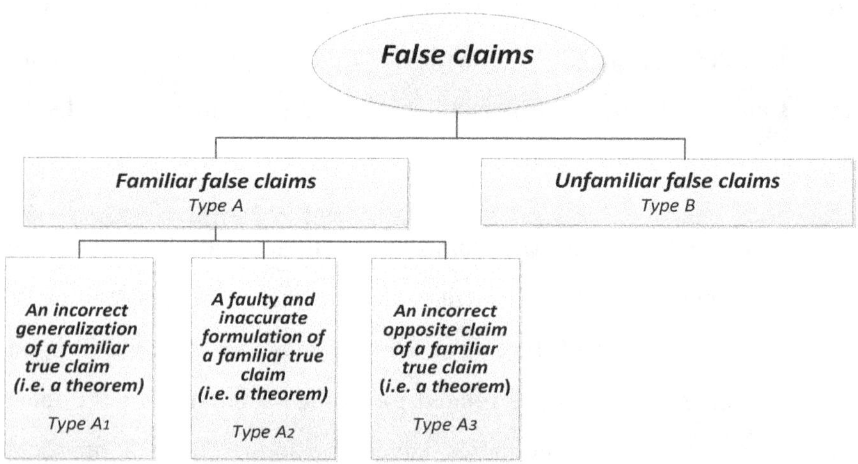

Figure No. 1: Types of false claims

Type A_1: An incorrect generalization of a "familiar" true claim, i.e. a theorem

An example of the cultivation and terms comprehension is included in the paper of Patkin and Plaksin (2011) that engages in congruent triangles. The claim "triangles whose sides are all equal (S. S. S) are congruent" is true. However, the truth of the claim (the corresponding equality of all the triangle sides leads to the congruency of the triangles) cannot be generalized to quadrilaterals. In the case of quadrilaterals, the claim "two quadrilaterals with correspondingly equal sides (S. S. S. S) are congruent" is false, since not every combination of major equal features is sufficient for congruency of polyhedrons (Plaksin & Patkin, 2016).

Type A_2: A faulty and inaccurate formulation of a "familiar" true claim

Formulating a true "familiar" claim in an inaccurate manner might lead to a situation, whereby the claim is false. Applying an accurate language is deemed highly important. Studies (Patkin 2011; Patkin & Gazit, 2011; Yopp, 2015) found that difficulties in mathematics stem not only from failure to acquire mathematics knowledge but also from the use of language in general and the abstract nature of mathematics language in particular. Moreover, they showed that changing the verbal formulation of a problem might entail a change in its mathematical characterization. As a result, this could prevent the successful solution of the problem. For instance: the congruency theorem relating to two triangles equal in two sides and one angle (S.S.A). If there is no indication that the angle is positioned opposite the long side, then in cases whereby the given angle is opposite the short side, the triangles are not necessarily congruent. Hence, as it is presented, the claim is false. Moreover, an inaccurate formulation of the "familiar" claim about an angle between a chord and a tangent and an inscribed angle, the following formulation: "An angle between a chord and a tangent is equal to the inscribed angle resting on that chord" leads to an error. The equality exists only if the inscribed angle resting on the same chord lies on the other side of the chord and not on the same side of the angle between the chord and a tangent.

Type A_3: An incorrect opposite claim of "familiar" true claim (i.e. a theorem) **(type A_3)**

The claim "the diagonals in the rectangles are equal" is a "familiar" true claim. The opposite claim: "If the diagonals of a quadrilateral are equal, the quadrilateral is a rectangle" is a false claim.

Pre-service Teachers' Knowledge of Geometry

During the three years of geometry studies, the 24 primary school pre-service teachers had to deal with true or false claims both in the lessons and tests. The tests sometimes consisted of claims all of which were true, false or partly true and partly false. In the questionnaire, the pre-service teachers were requested to respond to each claim, indicating whether it was true or false as well as justify and explain their determination. They were given no instruction as to the way of justifying and/or explaining. They could do it by a text, a drawing or a combination of both. The time allocated for filling in the questionnaire was 45 minutes, in order to prevent the option of guessing the answer.

The primary school pre-service teachers' responses to the ten false claims were analyzed according to the claim types: a "familiar" false claim and an "unfamiliar" false claim. Then, a reference was made to the types of arguments and the way of argumentation, while distinguishing between these two types of false claims.

Table No. 1 illustrates that the identification of the false claims ranged between 33%–100%. The successful identification percentage of four claims (claims 1, 2 and 5 of type A and claim 4 of type B) was high to very high (92%–100%). The percentage of successful identification of three additional claims (claim 3, 7 – type A and claim 8 – type B) was lower, ranging between 67%–75%. The pre-service teachers found it hard to identify three false claims: the "familiar" claim formulated in a faulty and inaccurate manner (claim 6, type A_2) as well as the two "unfamiliar" claims (Type B – claims 9, 10). The percentage of pre-service teachers who identified these claims as false was 33%–42%, the greatest difficulty being in type B (33%).

Table No. 1: Distribution of pre-service teachers' responses related to the identification of the claims and the types of reasoning

TYPE	**Pre-service teachers (N=24)**						
	Refutation of the claim (responded correctly)				**Identification of the claim as true (responded incorrectly)**		**No respns**
	Correct reasoning total	Incorrect reasoning total	No Reasoning total	**Total**	Incrrct./ Irrelvt	No Reason.	
A							
Claim 2	24	-	-	**24** (100%)	-	-	-
Claim 1	21	1	-	**22** (92%)	1	1	-
Claim 5	18	4	-	**22** (292%)	2	-	-
Claim 3	18	-	-	**18** (75%)	6	-	-
Claim 7	13	3	-	**16** (67%)	7	1	-
Claim 6	10	-	-	**10** (42%)	11	2	1
B							
Claim 4	23	-	-	**23** (96%)	1	-	-
Claim 8	13	4	-	**17** (71%)	6	1	-
Claim 9	5	3	1	**9** (38%)	1	10	4
Claim, 10	5	2	1	**8** (33%)	-	12	4
Total	**150**	**17**	**2**	**169**	**35**	**27**	**9**

Recommended Activities for Teaching False Claims as a Tool for Better Understanding

Examples of analysis of the reasoning given for the "familiar" false claims

Claim 2 was: "Every parallelogram can be inscribed in a circle". This claim is based on a false generalization of the true claim "Every triangle can be inscribed in a circle" or generalization of the true claim "Every rectangle can be inscribed in a circle". The claim was refuted by all the 24 pre-service teachers who justified their refutation. Thirteen pre-service teachers refuted

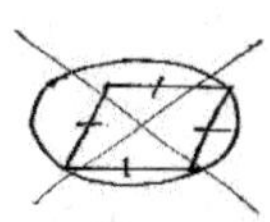

the claim, whereas nine pre-service teachers mentioned it was true in a parallelogram only if the sum of its opposite angles was 180°. Hence, this applied only to a rectangle and a square. The other four pre-service teachers added also a drawing of a rhombus or another parallelogram which is not a rectangle, showing that the claim was false. The remaining 11 pre-service teachers stipulated that there was no compliance with the condition of inscribing a quadrilateral into a circle, according to which the total opposite angles in a quadrilateral should be 180°.

Claim 1 was: "Every trapezoid can be inscribed [in a circle]". Twenty-one of the pre-service teachers refuted the claim and justified it correctly: "The total of opposite angles in an inscribed quadrilateral should be 180° and it exists only in an isosceles trapezoid, therefore the claim is false". The explanation given by the pre-service teachers was based on the theorem relating

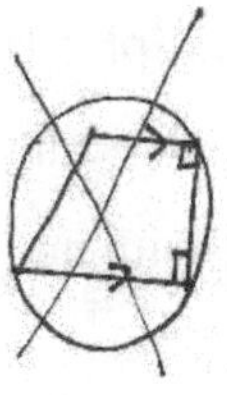

to conditions of a quadrilateral inscribed in a circle. Moreover, one pre-service teacher added to his reasoning a specific example, drawing a right-angled trapezoid that was not compatible with the claim.

Another pre-service teacher who refuted the claim gave an incorrect argument, drawing an isosceles trapezoid with two lower base vertices touching the circle and two others within the circle. This argument showed in fact that the trapezoid was drawn incorrectly and therefore it could not be inscribed. There might be another circle in which it would be inscribed but this is irrelevant to the refutation of the claim. Two pre-service teachers failed to refute the claim, indicating it was true. One of them pointed this out without reasoning and the other explained it by an incorrect argument, namely “every quadrilateral can be inscribed except for a rhombus and a parallelogram”.

Claim 5 was: “If the areas of two triangles are equal, the triangles are congruent”. In this case, identification of the claims as true or false was based on a logical connection of a true claim or claims that were “familiar”. It was

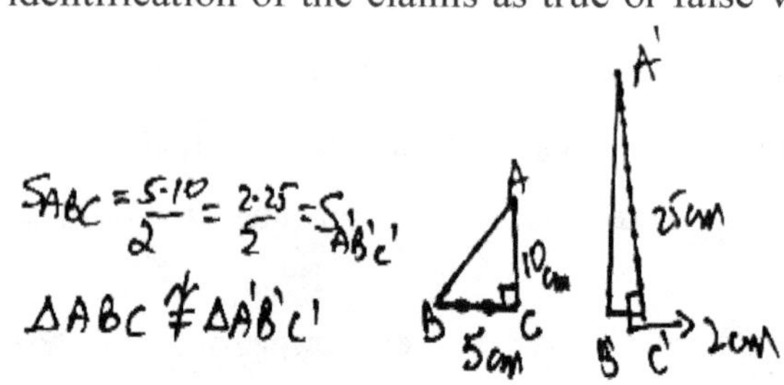

refuted by 22 pre-service teachers, 18 of them justifying their answer by giving correct arguments. Fourteen pre-service teachers presented counterexamples by text or drawing of particular cases. For example: “A right-angled triangle whose perpendiculars are 3 cm and 2 cm is not congruent with an isosceles triangle whose base is 2 cm and the altitude to the base is 3 cm" or "Two right-angled triangles, one right-angled triangle whose perpendiculars are 2 cm and 25 cm and one right-angled triangle whose perpendiculars are 10 cm and 5 cm, are not congruent". Two preservice teachers provided an argument and explained that "if the triangles are congruent one can conclude that the areas are equal but not the other way around. There might be two triangles whose sides are different but their areas are equal". Two other pre-service teachers related to the case of two triangles which were compatible with the condition that the altitude to the side in one triangle was equal to the length of the side of the other triangle and vice versa, the side of the other triangle was equal in length to the altitude of the other triangle. Four pre-service teachers who refuted the claim gave an incorrect argument such as: "It is not enough" (but didn't explain why). Two additional pre-service teachers identified the false claim as true, giving an incorrect and irrelevant argument, i.e. the formula of triangle area.

Claim 3 was: "The center of a circle circumscribing the triangle is always positioned inside the triangle". The false claim does not indicate that the triangle is acute-angled. Hence the error stemmed from the lack of a given or inaccuracy of the language and it was refuted by 18 pre-service teachers. Six pre-service teachers explained it by a drawing: five of them drew an obtuse-angled triangle as a counterexample with the center of the circle outside the triangle and one drew a right-angled triangle with the center of the circle positioned at the middle of the hypotenuse. An

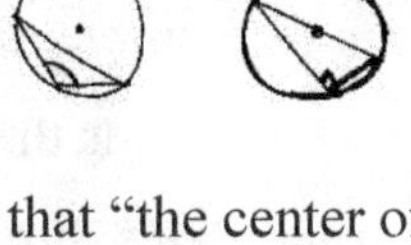

additional six pre-service teachers justified it and wrote that "the center of the circumscribing circle is the meeting point of the triangle perpendicular bisectors". The other six pre-service teachers justified the refutation of the claim by giving the explanation: "When in a right-angled triangle the meeting point is at the middle of the hypotenuse, in an obtuse-angled triangle the meeting angle is outside the triangle whereas in an acute-angled triangle it is within the triangle". Three of these pre-service teachers added drawings of the three above-mentioned possible situations. The remaining six pre-service teachers were wrong because they identified the claim as true, explaining it by an incorrect or irrelevant argument such as: "Every triangle has a circle".

Claim 7 was: "Two quadrilaterals whose sides are all proportional – are similar". This claim is a false generalization of the true claim: "Two triangles all the sides of which are proportional – are similar". The conclusion is that the theorem of similar triangles (S.S.S) cannot be generalized to quadrilaterals and hence this claim is also of type A_1.

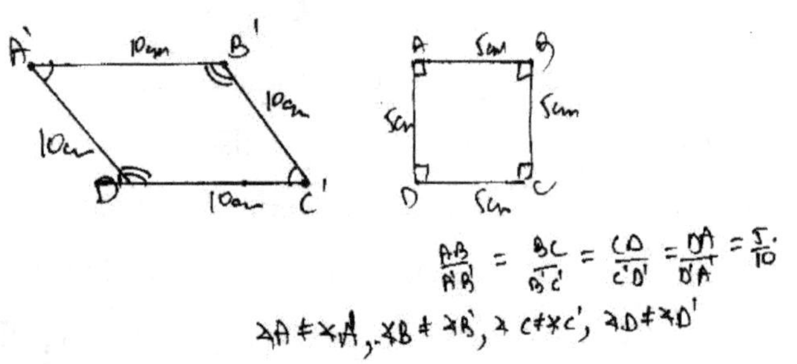

Sixteen pre-service teachers refuted the claim and justified their response. Four pre-service teachers explained by way of drawing a counterexample of a rhombus and a square which can be proportional in their sides but the inequality of their angles prevents them from being similar. Nine pre-service teachers gave the justification: "Insufficient proportion between their sides and equality between the angles is also required". The three other pre-service teachers whose arguments were false mentioned that equality in another angle was necessary but did not relate to the required position of that angle. Seven out of the eight pre-service teachers who failed to refute the claim grounded their incorrect response in the theorem of similar triangles due to the proportion between their sides, namely S.S.S. "Two quadrilaterals whose sides are all proportional – are similar". The eighth additional pre-service teacher identified the false claim as true, without giving any argument.

Claim 6 was: "An angle between a chord and a tangent is equal to the inscribed angle resting on that chord". This false claim based on the true claim is: "An angle confined between a tangent and a chord is equal to the inscribed angle resting on the other side of the same chord". The false claim did not indicate the position of the angle, namely, the error stemmed from the fact that a given was missing and the language used was inaccurate.

The success percentage of refuting this false claim out of the "familiar" false claims was the lowest. The formulation was faulty, ignoring the position of the inscribed angle. Only ten pre-service teachers (42%) pointed out that the claim was false and justified it. Seven among them justified its refutation by an accurate formulation of the claim, namely added the reference to the position of the inscribed angle which should be on the other side of the chord. The three other pre-service teachers added also a drawing that showed where the inscribed

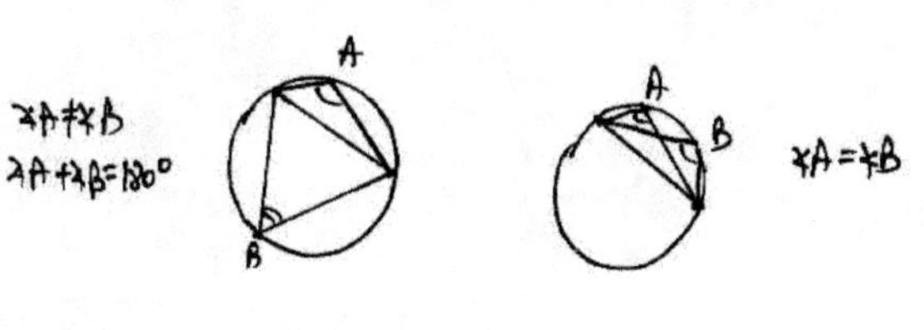

angle should be (on the other side of the chord). Thirteen pre-service teachers indicated that the (false) claim was true and one pre-service teacher did not respond to the question. Eight pre-service teachers among those who were incorrect and found that the claim was true, justified their determination by basing themselves on the accurate formulation of the claim which included the position of the angle on the other side of the chord. Those eight pre-service teachers failed to comprehend the meaning of the distinction between identification of a claim as false or true and the way of justifying that determination. They formulated the claim accurately in full and hence argued that it was true. Three other pre-service teachers incorrectly justified their false determination. Two additional pre-service teachers identified the false claim as true but gave no argument.

Analysis of the reasoning given for "unfamiliar" false claims relating to a true claim - Type B (claims 4, 8, 9, 10)

Claim 4 was: "Every two right-angled triangles are similar" and it was refuted by 23 pre-service teachers. This claim is not a generalization of a known theorem, is not an inaccurate formulation of a known theorem and is not an opposite claim of a true theorem and, therefore, it is a type B claim. The error in the identification stems from a false logical connection to a "familiar" true claim: "Two right-angled triangles that are equal also in one acute angle – are similar". Its refutation may be justified by relating to the theorem of congruency or by giving a counterexample.

Seven pre-service teachers refuted the claim by drawing a counterexample which related to a triangle whose angles were 60°, 90°, 30° versus a triangle whose angles were 90°, 45°, 45°. Four of these pre-service teachers presented the example by text, three others drew these triangles, indicating the size of the angles. Another pre-service teacher drew two right-angled triangles (one right-angled triangle with sides

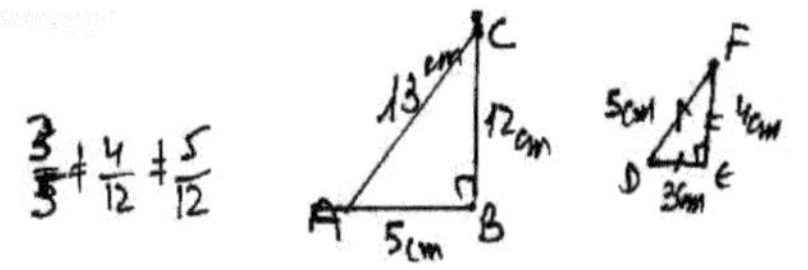

whose length was 3, 4, 5 cm and the second triangle with sides that were 5, 12,13 cm long). His reasoning was: "There was no proportion between the corresponding sides of the two triangles although both were right-angled. Hence, they were not similar". Fifteen pre-service teachers refuted the claim, relating to the fact that a right angle was insufficient and more equal angles were needed. Nine of these pre-service teachers refuted the claim by providing a general reasoning: "At least two right angles should be equal, one is insufficient", or "Not all the right-angled triangles are similar. Perhaps the equality exists only in the right angle and other acute angles are different". The other six justified their refutation of the claim by grounding themselves in different similar conditions. For example: "The claim is true if there is equality also in another acute angle or there is a proportion between one perpendicular and a hypotenuse of the two triangles, or a proportion between the two perpendiculars in the two triangles". That is, these pre-service teachers pointed out the need for equality of angles or sides which were compatible with the theorems of similar triangles - angle, angle, (A.A.), side, side, angle (S.S.A) or side, angle, side, (S.A.S). The remaining pre-service teacher was wrong because he identified the claim as true (explaining it by an incorrect and irrelevant argument).

Claim 8 was: “If a right-angled triangle and an obtuse-angled triangle do not have a common side, it is inconceivable that their areas are equal to each other”.

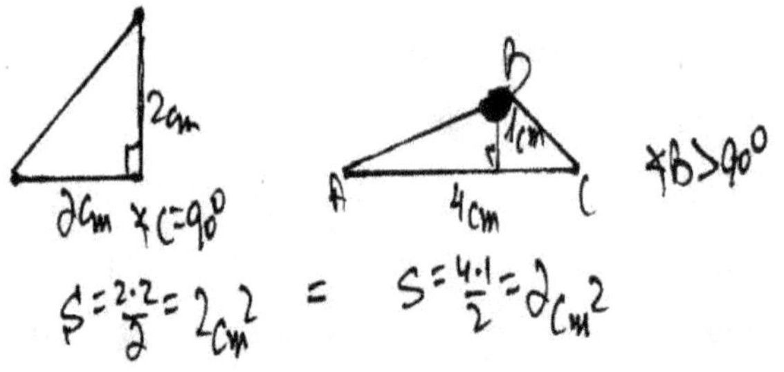

The claim engaged in unequal areas of triangles without a common side and was formulated in the negative. The mistaken identification of this claim also stems from a false logical connection to a “familiar” true claim: “Two different triangles (one right-angled and the other obtuse-angled) can have the same area even if they do not have a common side”. The formulation of the claim in the negative makes it more difficult to identify the claim as false. The difficulty in identifying the last two claims as false originates in a logical connection to the “familiar” true claim: “two triangles that are correspondingly equal in their sides are congruent”. Seventeen pre-service teachers refuted that claim, 13 of them presenting correct reasoning. Nine pre-service teachers refuted the claim by drawing two triangles, one right-angled and the other obtuse-angled and marked on them the values of the side lengths and the altitude. For example, a drawing of a right-angled and isosceles triangle with perpendiculars 2 cm long and an obtuse-angled triangle, one of its sides is 4 cm long and the altitude to this side is 1 cm. The other four pre-service teachers justified it in a general manner: "In the case of two triangles where the altitudes are equal and the sides to which the altitude passes are equal – regardless of the triangle type – their areas are equal". Four pre-service teachers who refuted the claim gave an incorrect argument. Seven additional pre-service teachers identified the false claim as true. Six out of them explained it by an incorrect and irrelevant argument. The remaining pre-service teacher did not respond to this claim.

Claim 9 engaged in two quadrilaterals, each consisting of two triangles so that both triangles in one quadrilateral were congruent with the two triangles in the second quadrilateral. The pre-service teachers were asked whether the quadrilaterals were congruent and if yes – was it because all their sides and all their angles were correspondingly equal. Alternately, if the pre-service teachers maintained that the quadrilaterals were not congruent, they were asked to draw a counterexample. The problematic of claim 9 resides in identifying the fact that even if the triangles are congruent and they form quadrilaterals consisting of congruent triangles, the quadrilaterals are not congruent because the order of the sides is not identical (once a kite is obtained and once a parallelogram).

The findings showed that nine of the pre-service teachers (37%) identified the claim as false. Five among them (21%) gave correct reasoning and drew as counterexample two different-sided triangles which, when connected by the equal side, created non-congruent quadrilaterals. For example, one pre-service teacher drew two congruent triangles whose sides were 5, 6, 7 cm long. The side of the quadrilaterals which could have been created were: 5, 6, 6, 5 cm or 5, 6, 5, 6 cm. In both of them, there was a diagonal which was the side in the triangles equal in length to 7 cm, but these quadrilaterals were not congruent. One quadrilateral was a kite and the other a parallelogram. Three other pre-service teachers gave an incorrect reasoning and the ninth pre-service teacher didn't give any argument. Eleven pre-service teachers (46%) failed to identify that the claim was false and incorrectly maintained that the quadrilaterals were congruent. One of these pre-service teachers justified her determination that the claim was true by stating that "the triangles were congruent!!! The angles and sides of each triangle were correspondingly equal" (underscored by the pre-service teacher). The remaining four pre-service teachers did not respond to this claim.

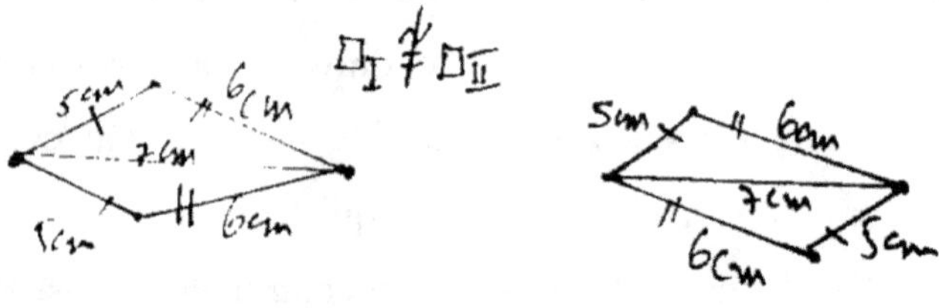

Claim 10 comprised a drawing with two hexagons equal in six sides and six angles.

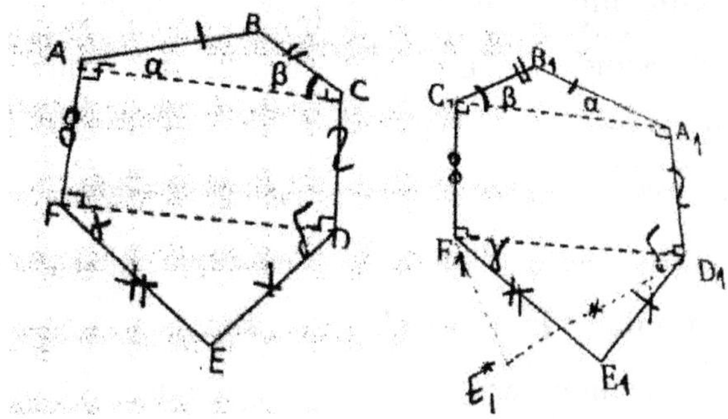

However, while the corresponding sides were equal, the angles were not correspondingly equal. The pre-service teachers were requested to identify whether the hexagons were congruent and, if yes, was it due to the fact that all their sides were correspondingly equal and all the angles correspondingly equal. Alternately, if the pre-service teachers maintained that the hexagons were not congruent, they had to justify their determination. Eight pre-service teachers (33%) identified that the claim was false and that the hexagons were not congruent. Five of them (21%) said: "one has to pay attention that the sides are correspondingly equal but are not correspondingly positioned". Two pre-service teachers gave an incorrect explanation that "the sides were correspondingly equal and the angles too were correspondingly equal" rather than "the two angles were not correspondingly equal". The eighth pre-service teacher did not justify his determination. Twelve additional pre-service teachers identified the false claim as true but gave no argument. Four pre-service teachers did not respond to this claim.

To sum up, all the pre-service teachers responded to seven of the ten claims: five out of six (83%) “familiar” false claim (type A) and two out of four (50%) “unfamiliar” false claims (type B). On the other hand, some pre-service teachers didn’t respond and didn’t identify three claims: one “familiar” false claim (type A_2) and two “unfamiliar” false claims (type B). More than 45% of the pre-service teachers (46%–54%) found it difficult to identify these claims and, therefore, their response was incorrect. The findings indicate also a difference in the pre-service teachers’ answers to “familiar” and “unfamiliar” false claims as presented in Table No. 2 (distribution of the answers related to the identification of “familiar” false claims (type A) vis-à-vis the distribution of the answers related to “unfamiliar” false claims (type B).

Table No. 2: No. of pre-service teachers who identified "unfamiliar" false claims and "familiar" false claims

Correctly identified ffc* (type A)	**No. of pre-service teachers who correctly identified "unfamiliar" false claims (type B)**					
	Four Uffc (100%)**	**Three Uffc** (75%)**	**Two uffc** (50%)**	**One uffc** (25%)**	**Incorrect Identifi-cation**	**Total**
Three fcc* (50%)	-	-	4 (2 jstfd 2 uffc** 3 jstfd 1 uffc**)	-	-	**4**
Four fcc* (67%)	1 (jstfd 2 uffc**)	-	3 (jstfd 1 uffc**)	-	-	**4**
Five fcc* (83%)	3 (jstfd 3 fffc**)	4 (2 jstfd 3 uffc** 2 jstfd 1 uffc**)	3 (2 jstfd 2 uffc** 1 jstfd 1 uffc**)	-	1	**11**
All the six ffc* (100%)	1 (jstfd 3 fffc**)	3 (1 jstfd 3 uffc** 2 jstfd 2 uffc**)	1 (jstfd 2 uffc**(	-	-	**5**
Total	**5 (4 jstfd 3 uffc** 1 jstfd 2 uffc**)**	**7 (3 jstfd 3 uffc** 2 jstfd 2 uffc** 2 jstfd 1 uffc**)**	**11 (5 jstfd 2 uffc** 6 jstfd 1 uffc**)**	**-**	**1**	**24**

*fcc – familiar false claims; **uffc – unfamiliar false claims

This chapter compared the successful identification of "familiar" false claims (type A) and of "unfamiliar" false claims (type B). Table No. 2 illustrates that not all the pre-service teachers who succeeded to refute "familiar" claims demonstrated the same level of success and deductive reasoning regarding "unfamiliar" claims. Only four out of 16 pre-service teachers (25%) who correctly identified 83%–100% "familiar" claims (namely five or six claims as false) and justified part or all the refutations, also identified the four "unfamiliar" claims as false. Another pre-service teacher who correctly identified only four "familiar" claims as false (67%) identified the four "unfamiliar' claims as false. Seven out of 16 pre-service teachers (44%) who correctly identified 83%–100% "familiar" claims identified three claims of type B ("unfamiliar" claims") as false. The other eleven pre-service teachers identified only two "unfamiliar" claims as false. Four of them failed to identify 50% of the "familiar" claims, six pre-service teachers succeeded to identify four or five "familiar" claims as false (67%–83%) and only one succeeded to identify all the six "familiar" claims as false. Five of these eleven pre-service teachers succeeded to justify the identification of claim 8 as false. In addition, all the elven pre-service teachers failed to justify claim 9 and claim 10. It is important to mention that all the pre-service teachers except one managed to identify claim 4 as false and to justify the refutation. This pre-service teacher failed in the identification of all the three other "unfamiliar" claims too, even though he succeeded to identify correctly five "familiar" claims as false. That is, the findings indicated the pre-service teachers' difficulties in identifying most of the claims which were "unfamiliar" to them.

The findings illustrate that pre-service teachers who managed to identify and refute the "unfamiliar" claims as false succeeded also in identifying and refuting the "familiar" claims. Conversely, pre-service teachers who failed to identify "familiar" claims as false failed also to identify the "unfamiliar" claims as false. Moreover, the findings show that in spite of correctly identifying "familiar" claims as false, some pre-service teachers failed to identify "unfamiliar" claims as false. It is noteworthy that all the pre-service teachers who succeeded to identify claims 9 or 10 as false, succeeded to identify claim 6 as false.

Discussion, Conclusions, and Recommendations

This teaching experience has resulted in the idea of exploring whether primary school mathematics pre-service teachers can identify false claims whose formulation 'reminded' them of one of the previously studied theorems. The writer of this chapter has also decided to investigate whether these pre-service teachers (towards the end of their third academic year, and after having attended three annual disciplinary courses in geometry) were capable of identifying types and sub-types of false claims and justify their refutation.

The research findings illustrate that success, as well as the failure to identify false claims, was demonstrated in both claim types, "familiar" and "unfamiliar". The lack of attention and awareness of meticulousness and accuracy of the language can account for this. In some cases, the pre-service teachers disregarded the accuracy of the language, a highly important factor for coping with mathematical problem solution (Patkin & Gazit 2011; Yopp, 2015). However, their reasoning indicated they understood something was missing and they added that ‘something’ in the reasoning. Adding the missing detail was for them a reasoning that the claim ‘became true’. They even underscored the missing detail.

Furthermore, two of the “unfamiliar” claims referred to the term “correspondingly” which is a difficult concept for the pre-service teachers. Hence, it is recommended by Patkin (2011), and Plaksin and Patkin (2016) to pay more attention to the concept of correspondence and to the integration of claims that are less discussed within the framework of such courses.

The findings analysis also illustrated that the primary school pre-service teachers justified their drawings by adding to some of them a text explaining the drawings. This finding is similar to the findings of Zeybek (2016). He maintains that there are pre-service teachers with deductive reasoning who settle for one counterexample of a general example nature. On the other hand, those pre-service teachers who had not yet attained deductive reasoning or encountered difficulties, justified the refutation of the claims by means of specific examples. These findings are in line with the study of Messinger (2006) and Patkin (2012), who referred to this way of reasoning as ‘over-deductivity’.

Another interesting finding was a numerical counterexample that was accompanied by the ‘translation’ of the drawing into text. Using numerical counterexamples which constitute only a known prototype, can indicate the pre-service teachers’ lack of confidence and, therefore, they only hold on to something “familiar”. These false justifications led to the conclusion that they were made without thinking and in a technical manner, i.e. in certain cases the reasoning contradicted the answer related to the identification of the claim as false.

Patkin (2012) indicates that deductive reasoning requires a high level of thinking. Hence, only when learners attain such a level of thinking, are they capable of understanding proofs and of proving claims independently. In this chapter, the classification of the claims as false and the reasoning thereof require a high level of thinking. Perhaps some of the pre-service teachers who identified claims as false and justified their determination had not developed sufficient deductive reasoning. Thus, they correctly responded to the questions due to the memorization of the material rather than to the high level of mathematical thinking. Their argument ("if the triangles are congruent, the areas are equal but not vice versa, maybe the sides are different and the areas are equal") entailed hesitations about the pre-service teachers' true level of understanding. The provision of this argument might constitute the memorization of the material rather than the understanding thereof. The broader issue of pre-service teachers' ability to develop a deep understanding and reasoning ability and proving or disproving mathematical statements, especially in deductive geometry, is important to their future experience as mathematics teachers. This helps them to promote the same understanding and abilities of their learners, a process that affects the learners' achievements in mathematics.

This chapter describes the way primary school pre-service teachers cope with the refutation of "familiar" false claims of three sub-types and "unfamiliar" false claims and herein resides its innovation. The assumption was that the identification of subtypes of false claims might help planning a designated training program for teaching pre-service teachers. However, the findings illustrated that the difference of successful refutation between these two types of claims (type A and type B and the three subtypes) was not so crucial. It was rather the accuracy in a mathematical language that accounted for this difference. The analysis of the false claim types may lead teacher-educators as well as school teachers to integrate into their lessons a larger number of claims which are false due to inaccurate, more complex and unknown formulation. The research findings indicate that the ability to refute "unfamiliar" false claims (formulated with several terms which are less discussed in the syllabus) comprises mindfulness to the accurate formulation of the "familiar" theorems. Consequently, it is recommended developing learners' awareness of accurate formulations and careful attention to unsubstantiated generalizations for the purpose of enhancing the deductive thinking of those learners.

Nevertheless, in spite of the support of the above conclusions, the findings should be taken with some reservation, considering the small number of participants. It is therefore recommended conducting a similar study of a larger pre-service teacher group for checking whether the findings are similar. Moreover, other studies should be conducted of pre-service teachers at the beginning of their studies as well as throughout their three years at the college, with the purpose of exploring whether there has been an improvement in their ability to identify false claims of various types and justify their determination. These studies should investigate whether there are differences in the participants' ability to use an accurate language, a factor that might promote the learners' ability to refute claims and cope with problems formulated in a complex manner. Such problems require expanding the ability to identify whether claims are true or false on the basis of a logical connection. In addition, these courses would enable the pre-service teachers to better differentiate between correct identification and refutation of false claims in geometry and to enhance their deductive reasoning ability.

This chapter contributes to knowledge in this field by allowing educational practitioners to classify the arguments according to the categories defined in this chapter. Furthermore, educators would be able to learn about the learners' ability to presents various types of arguments while using accurate mathematical language, i.e. appropriate theorems, definitions, and drawings.

References

1. Balacheff, N. (1991). The benefits and limits of social interaction: The case of mathematical proof. In A. J. Bishop, S. Mellim-Olsen & J. Van Dormolen (Eds.). *Mathematical Knowledge: Its growth through teaching* (pp. 175–194). Dordrecht, The Netherland: Kluwer Academic Publishers.
2. Barkai, R. (2009). *Mathematical and didactical aspects of proving mathematical statements*. Thesis submitted for the degree of Doctor of Philosophy. Tel-Aviv, Israel: Tel-Aviv University. [Hebrew]
3. Barkai, R., Tsamir, P., Tirosh, D., & Dreyfus, T. (2002). Proving or refuting arithmetic claims: The case of elementary school teachers. *Proceedings of the 26th International Conference on the Psychology of Mathematics Education*, *2*, 57–64. Norwich, UK: University of East Anglia.

4. Chazan, D. (1993). High school geometry students' justification for their views of empirical evidence and mathematical proof. *Educational Studies in Mathematics, 24*(4), 359–387.
5. Clements, D. H., & Battista, M. (1995). Geometry and proof. *Mathematics Teacher, 88*, 48–54.
6. Dreyfus, T. (1999). Why Johnny can't prove. *Educational studies in mathematics 38*(1–3), 85–109.
7. Eco, U. (1983). Homs, hooves, insteps: some hypotheses on three types of abduction. In U. Eco & T. Sebeok (Eds.), *The sign of three: Duplin, Holmes, Peirce* (pp. 198–220). Bloomington, IN: Indiana University Press.
8. Fischbein, E. (1982). *Intuition and proof. For the learning of mathematics, 3,* 9–18. Montreal, Quebec. Canada: FLM Publishing Association.
9. Fischbein, E., & Kedem, I. (1982). Proof and certitude in the development of Mathematical thinking. In A. Vemandel (Ed.). *Proceeding of the sixth International Conference for the Psychology of Mathematics Education* (pp. 128–131). Antwerp, Belgium: Universitaire Instelling Antwerpen.
10. Furinghetti, F., Olivero, F., & Paola, D. (2001). Students approaching through conjectures: Snapshots in classrooms. *International Journal of Mathematics Education in Science and Technology, 32*(3), 319–335.
11. Hanna, G. (2000). Proof, explanation and exploration: An overview. *Educational Studies in Mathematics, 44*(1–2), 5–23.
12. Herbst, P. G. (2002). Engaging students in proving: A double bind on the teacher. *Journal for Research in Mathematics Education, 33*(3), 176–203.
13. Johnson, D. W., & Johnson, R. T. (2009). Energizing learning: The instructional power of conflicts. *Educational Research 38*(1), 37–51
14. Jones, K. (2000). The students experience of mathematical proof at university level. *International Journal of Mathematical Education in Science and Technology, 31*(1), 53–60.
15. Lakatos, I. (1976). *Proofs and refutations*. Cambridge: Cambridge University Press.
16. Lin, F. L. (2005). Modeling students' learning on mathematical proof and refutation. In H. L. Chick & J. L. Vincent (Eds.). *Proceeding of the 29th conference of PME* (pp. 3–18). Melbourne.

17. Messinger, R. (2006). *How do students prove and refute algebraic claims of the type "every..." and "there exists...": The case of the deductive approach.* MA dissertation, Tel-Aviv, Israel: Tel-Aviv University. [Hebrew]
18. National Council of Teachers of Mathematics (NCTM). (2000). *Principles and Standards for School Mathematics*. Reston, VA: NCTM.
19. Patkin, D. (2011). The interplay of language and mathematics. *Pythagoras - Journal for Research in Mathematics Education, 32*(2), 1–7.
20. Patkin, D. (2012). High school students' perceptions of geometrical proofs. Proving and refuting geometrical claims of the "for every..." and "there exists" type. *International Journal of Mathematical Education in Science and Technology, 43*(8), 985–998.
21. Patkin, D., & Gazit, A. (2011). Effect of difference in word formulation and mathematical characteristics of story problems on mathematics preservice teachers and practicing teachers. *International Journal of Mathematical Education in Science and Technology, 42*(1), 75–87.
22. Patkin, D., & Plaksin, O. (2011). Congruent Triangles. Sufficient conditions and insufficient conditions. Suggested milestones for inquiry and discussion. *Research in Mathematical Education, 15(*4), 327–340.
23. Peirce, C. S. (1960). *Collected papers.* Cambridge, MA: Harvard University Press.
24. Plaksin, O., & Patkin, D. (2016). Exploring the minimum conditions for congruency of polygons. *Learning and Teaching Mathematics, 21,* 42–46.
25. Potari, D., Zachariades, T., & Zaslavsky, O. (2010). Mathematics teachers' reasoning for refuting students' invalid claims. In: V. Durand-Guerrier, S. Soury-Lavergne & F. Arzarello (Eds.). *Proceedings of CERME, 6,* 281–290.
26. Vinner, S. (1983). The notion of proof – some aspects of students' views at the senior high level. In R. Hershkowitz (Ed.). *Proceeding of the 7th Conference of the International Group for the Psychology of Mathematics Education* (pp. 289–294). Rehovot, Israel: The Weizmann Institute of Science.
27. Vinner, S. (1997). The pseudo-conceptual and the pseudo-analytic thought processes in mathematics learning. *Educational Studies in Mathematics, 34*(2), 97–129.
28. Waring, S. (2000). *Can you prove? Developing concepts of* proof *in primary and secondary schools*. Leicester, UK: The Mathematical Association.

29. Weber, K. (2002). Beyond proving and explaining: Proofs that justify the use of definitions and axiomatic structures and proofs that illustrate technique. *For the Learning of Mathematics, 22*(3), 14–17.
30. Weber, K. (2008). How mathematicians determine if an argument is a valid proof. *Journal for Research in Mathematics Education, 39,* 431–459.
31. Wohlgemuth, A. (2003). *Deductive Mathematics - an introduction to proof and discovery.* Syracuse, NY: Andrew Wohlgemuth.
32. Yackel, E., & Hanna, G. (2003). Reasoning and proof. In J. Kilpatrick, W. G. Martin & D. Shiffer (Eds.). *A research companion to principles and standards for school mathematics* (pp. 227–236). Reston, VA: NCTM.
33. Yollesh, N., & Dreyfus, T. (2003). Does a proof render the truth of a claim decidable? In A. Rogerson (Ed.). *Proceeding of the International Conference on the Decidable and the Undecidable in mathematics education* (pp. 68–71). Brno, Czech Republic.
34. Yopp, D. A. (2015). Prospective elementary teachers' claiming in responses to false generalizations. *The Journal of Mathematical Behavior, 39*, 79–99.
35. Zaslavsky, O., & Ron, G. (1998). Students' understandings of the role of counterexamples. In A. Olivier, & K. Newstead (Eds.), *Proceedings of the 22 Conference of the international Group for the Psychology of Mathematics Education* (Vol. 4, pp. 225–232). Stellenbosch, South Africa: University of Stellenbosch.
36. Zazkis, R., & Chernoff, E. J. (2008). What makes a counterexample exemplary? *Educational Studies in Mathematics, 68*, 195–208.
37. Zeybek, Z. (2016). Pre-service elementary teachers' conceptions of counterexamples. *International Journal of Education in Mathematics, Science and Technology (IJEMST), 5*(4), 295–316. DOI:10.18404/ijemst.70986.

Appendix A

Below are claims in geometry. You have to identify whether the claims are true or false and justify it in any way you choose

1. **Claim 1** was: "Every trapezoid can be inscribed in a circle".
   (Yes/No)
   argument:
   ____

2. **Claim 2** was: "Every parallelogram can be inscribed in a circle".
   (Yes/No)
   argument:
   ____

3. **Claim 3** was: "The center of a circle circumscribing the triangle is always positioned inside the triangle".
   (Yes/No)
   argument:
   ____

4. **Claim 4** was: "Every two right-angled triangles are similar".
   (Yes/No)
   argument:
   ____

5. **Claim 5** was: "If the areas of two triangles are equal, the triangles are congruent".
   (Yes/No)
   argument:
   ____

6. **Claim 6** was: “An angle between a chord and a tangent is equal to the inscribed angle resting on that chord”.

   (Yes/No)

   argument:

   ____

7. **Claim 7** was: “Two quadrilaterals whose sides are all proportional – are similar”.

   (Yes/No)

   argument:

   ____

8. **Claim 8** was: “If a right-angled triangle and an obtuse-angled triangle do not have a common side, it is inconceivable that their areas are equal to each other”.

   (Yes/No)

   argument:

   ____

9. Two quadrilaterals ABCD and EFGH are given. The diagonal AC creates with the sides of the quadrilateral ABCD two triangles ABC and ADC and diagonal EG creates with the sides of the quadrilateral EFGH two triangles EFG and EHG. Thus, the triangles ADC and EFG are congruent and triangles EHG and ABC are also congruent. Are the quadrilaterals ABCD and EFGH congruent quadrilaterals? Please choose the correct answer from the following answers:

a. **Yes, the quadrilaterals are congruent** because all their sides are correspondingly equal and all their angles are correspondingly equal.

b. **The quadrilaterals are not necessarily congruent** (if you chose this answer, **please draw a counterexample**.

10. Two hexagons are given

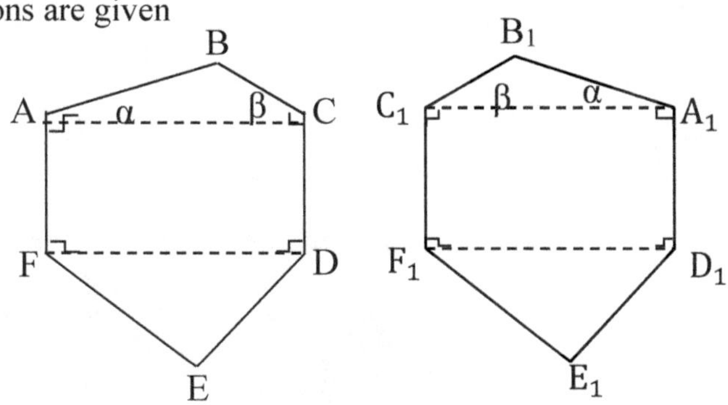

$AB = A_1B_1, BC = B_1C_1, CD = A_1D_1, DE = D_1E_1, EF = E_1F_1, AF = C_1F_1$

$\measuredangle ABC = \measuredangle A_1B_1C_1,\ \measuredangle BCD = \measuredangle B_1C_1F_1,\ \measuredangle CDE = \measuredangle A_1D_1E_1$

$\measuredangle DEF = \measuredangle D_1E_1F_1,\ \measuredangle EFA = \measuredangle E_1F_1C_1,\ \measuredangle FAB = \measuredangle D_1A_1B_1$

Are the hexagons congruent? Please choose the correct answer from the following answers:

a. **The hexagons are congruent** because all their sides are correspondingly equal and all their angles are correspondingly equal.

b. **The quadrilaterals are not congruent** (if you chose this answer, **please provide a short argument.**

https://doi.org/10.1142/9789811218750_0007

Chapter 7

Developing Mathematical Creativity through Problem Posing

Atara Shriki

Introduction

In the school reality, most of the mathematical problems are presented by the teachers or the textbooks. Nevertheless, as indicated in Chapter 1 of this book, the 'real' mathematical practice, illustrated by the works of leading mathematicians, is motivated by problems which they themselves have posed. Solving problems and posing problems do in fact constitute two of the key features of the creative mathematical practice (Silver, 1997). Yet, many of the problems included in the textbooks end with instructions, such as 'prove that…' or 'find the…'. That is, using a formulation, whereby the conclusion stemming from the data is explicit or a formulation that narrows learners' ability to see the whole picture of the problem. Obviously, when learners are requested to prove a certain mathematical statement, then a priori they are confident in its truth. Thus, they do not have the opportunity to make hypotheses regarding the different options encompassed in the situation of the problem presented to them, as well as the opportunity to guide the inquiry activity by themselves. Consequently, learners are denied the opportunity to develop their own creativity, and they frequently find themselves in a situation where they try finding a proof merely for 'pleasing' the teacher. Often, learners are not interested in coping with problems of this type, and as a result, they develop a wrong image of mathematics, according to which this is a closed, arbitrary and technical discipline (Shriki, 2007). Furthermore, we often realize that learners do not comprehend the meaning of a certain problem and are unaware of the characteristics of the logical relations underlying its data, even after having successfully coped with it. This is mainly due to the fact that the very engagement in a problem, is done in a mechanical and algorithmic manner, without enabling the learners to pose problems that are relevant to the given situation (Shriki, 2004). Conversely, when learners pose the problems, their mathematics knowledge is enhanced, since, in order to pose a valid problem, they have to explore the covert and implicit relations embedded in the situation (Lavy & Shriki, 2010; Shriki & Lavy, 2012). Moreover, because problem solving and problem posing oblige learners to apply associated cognitive actions, then posing problems have a positive effect on their ability to solve problems (Nikata & Shimada, 2005; Silver & Cai, 1996). In addition to the advantages of engaging in problem posing from the cognitive aspect, there are advantages from the emotional aspect as well. When learners pose problems that interest them, it naturally develops their inner motivation to engage in these problems (Shriki, 2004). By doing that, they assume part of the responsibility for the learning, while developing skills of independent learners (English, 1997). The process of posing problem also evokes the learners' sense of ownership on the learning material, reinforcing their inquisitiveness and enthusiasm about learning mathematics (Cunningham, 2004).

'What If Not?' – Posing Problems Strategy

The relation between positing problems for the purpose of developing creativity, reinforcing cognitive ability and enhancing knowledge and emotional aspects, indicates the importance of engaging in problem posing as an inseparable part of the learning process as a whole. Hence, the question is how to develop learning environments in which posing problems is intertwined in a natural manner, without burdening teachers' work. This sub-chapter presents one of the effective ways of developing learners' ability to pose new problems on the basis of a given problem, namely the 'What If Not?' strategy.

In order to assist learners in posing problems, Silver (1997) suggests implementing the approach developed by Brown and Walters (1969, 1990), frequently referred to as 'What-If-Not?' (abb. WIN). This strategy is grounded in the idea that changing the components of a given problem ('the starting point'), might lead to the posing of thought-evoking problems. As a result, learners will engage in inquiry and uncovering of mathematical regularities. This problem posing strategy leads learners through three consecutive levels, starting with a renewed examination of the given problem. At the first stage (Level 1), learners are requested to generate a list of properties or attributes included in the problem. At the second stage (Level 2), the learners focus on each of the attributes specified on the list, raise the WIN question, suggesting alternative attributes. The third stage (Level 3) consists of posing new problems and questions, based on the alternatives proposed in the second stage. Following this path, step by step, supports the development of learners' new insights. This is due to the fact that '*Only after we have looked at something, not as it 'is' but as it is turned inside out or upside down, do we see its essence or significance*' (Brown & Walter, 1990, p. 15).

Implementing the WIN strategy allows teachers to easily expand their teaching repertoire, since the given problems can (and even should) be taken from the textbooks. Thus, for example, instead of coping with five different problems that are not connected in any way, learners engage in five problems that maintain internal relationships. This does not only avoid damaging the sequence of progress in the curriculum, but also attributes an added value to the learning, since engaging in problems with internal logical relations enables learners to better comprehend the meaning of the given problem, instead of focusing exclusively on its solution. Accordingly, teachers' role is to offer learners situations with the potential of posing interesting questions. Moreover, they have to create a learning environment, enabling learners to cope with problems of real interest, inquire about them and engage in their solution (Shriki, 2007). It is noteworthy that teachers too benefit from such a learning environment. Learners frequently pose problems that might challenge and interest the teachers, problems the solution of which leads the teachers themselves to uncover new relations and properties with which they have not been acquainted.

Below is an example of employing the WIN strategy based on a simple given problem. But first we make two comments:

- Stoyanova and Ellerton (1996) classified the situations of problem posing into three types: free situation, semi-structured situation and structured situation. Problem posing situation will be referred to as free "*when students are asked to generate a problem from a given, contrived or naturalistic situation. Some directions may be given to prompt certain specific actions*" (p. 519); as semi-structured "*when students are given an open situation and are invited to explore the structure and to complete it by applying knowledge, skills, concepts, and relationships from their previous mathematical experiences*"; and as structured "*when problem-posing activities are based on a specific problem*" (p. 520). In her articles, Ellerton (1986a, 1986b) presents several examples of each of the situations. In the example of a free situation of problem posing, she asked the learners to come up with mathematical problems. She instructed them to formulate a problem that their peers would find difficult to solve. Furthermore, she asked them to write a letter to a friend who was ill, telling him or her about the mathematics which they have studied during the last three weeks. As part of the letter, they had to write questions like the ones they came across during that period. Stoyanova and Ellerton (1996) requested the learners to formulate problems in which they had to relate also to those who have to solve them. For example, the learners were required to formulate problems of three types for a mathematics competition: problems that they like, problems that they think their friends would have difficulties to solve and problems on a difficulty level suitable to their mathematics teacher. The example of a semi-structured situation of problem posing presented a problem with only partial information (e.g. a drawing of a right-angled triangle) and the learners had to pose as many relevant questions as possible. Another request was to pose as many problems as possible for the following calculation: $3 \cdot 25 + 15:5 - 4$. Another option was asking the learners to formulate a problem whose solution was given. In the examples of a structured situation of problem posing, Stoyanova and Ellerton (1996) presented three possible situations, designed to help the learners understand a certain problem and the structure of its solution. Then, they had to investigate possible internal relations between the problem formulation and its optional solutions. In some problems, only the situation was described and the question was omitted. In this case, the learners had to pose a series of possible problems and arrange them in an appropriate order. In other cases, they were required to add attributes to the problem structure, raise questions or find an unnecessary information in the problem, formulating it again. My personal experience (Shriki, 2013; Shriki & Lavy, 2012) illustrates that teachers believe learners' initial experience of implementing the WIN strategy, should be in a structured situation, that is based on a specific problem, well-defined, and complete from the point of view of the problem data. The reason is that, in such a situation, the given problem provides the learners with a 'safe anchor to hang on', rendering it easier for them to learn about the various aspects of problem posing. Therefore, the examples given below are based on a problem presented in a structured situation. Please note that all the presented examples are taken from the various chapters of this book.

- Chapter 2 of this book illustrates that mathematics teachers are more willing to implement learning processes associated with the development of creativity when geometry teaching is concerned (Levav-Waynberg & Leikin, 2012). This is due to the fact that geometric problems are perceived as problems that do not lead to algorithmic or procedural solutions, unlike other topics of school mathematics. Consequently, many teachers do not view the solution of geometric problems in different ways as a task that comes at the expense of the time allocated to the teaching of the curriculum. Based on the above, we are going to demonstrate the problems posing process based on a given geometric problem. We hope that this will facilitate teachers' comprehension of the scope of the potential embodied in this approach and, thus, will increase their willingness to embrace it.

The Given Problem

For the purpose of demonstrating the WIN strategy, we have chosen as the original problem the problem presented in Chapter 3 of this book (Patkin & Levenberg: "Creative approach to finding the sum of internal and external angles of convex polygons"). Instead of formulating the original problem in the following way: "A certain triangle is given. What is the sum of the triangle internal angles?", the formulation of the given problem has been changed to: "An equilateral triangle is given. What is the sum of the triangle internal angles?". The minor change in the problem formulation is designed to allow learners to experience the surprise that comes with the gradual discovery that the sum of angles in *every* triangle is fixed, regardless of the triangle type or size.

Comments:

- It is recommended preparing in advance equilateral triangles of different sizes. Thus, learners will encounter no difficulty in drawing the triangle. Moreover, they will be able to get an impression already at this stage that the size of the triangle has no effect on the sum of its angles.

- Since the formulation of the question above may imply that this concerns a fixed sum of sizes ("what is the sum of the triangle internal angles?"), the formulation of the question can be slightly changed. This is particularly valid if we prepare in advance clippings of triangles and distribute them to the learners. In such a case, the problem can be formulated in the following way: "An equilateral triangle is given. What is the sum of the internal angles of the triangle that you have before you?"
- When you ask the learners explicitly: "What is the sum of the triangle internal angles?" they are explicitly directed to the issue on which they have to focus. Teachers interested in making the situation more open can settle for distributing the triangle clippings to the learners, enabling them to raise questions associated with the given situation, without limiting a priori the directions of the inquiry.

Solution:

As proposed in the paper of Patkin and Levenberg (Chapter 3 of this book), learners can be instructed to cut the triangle angles and place them around one of the vertices. The objective is to realize that the sum of the angles is equal to the size of a straight angle, namely 180°.

Now we will demonstrate the process of posing problems on the basis of this given problem, by applying the WIN strategy.

a. Level 1: Making a list of the problem attributes

1. A polygon is given; 2. The polygon has three sides; 3. The three sides are equal in length; 4. The sum of the angle degrees should be calculated; 5. The angles are internal to the triangle.

This list is only an example and it can obviously be reduced or expanded. The level of detail depends on the learners' previous knowledge. However, the more detailed the list, the greater the option of posing diversified problems.

b. Level 2: Posing the WIN question and suggesting alternative attributes

With the sign "~" we mark the negation of a certain property. Thus, for example, the meaning of (~1) is the negation of attribute 1. The sign $(\sim 1)_1$ implies 'alternative 1 to attribute 1'. The alternatives proposed, later on, are merely examples and the list can be expanded.

A comment:

- At the stage of problem posing, it is recommended not limiting the learners to a certain type of problems, in order to allow them to develop their mathematical creativity in a 'free' manner. This can result in a situation, whereby some of the raised questions have no meaning from a mathematical point of view (see for example $(\sim 4)_1$ below). However, it is suggested postponing the process of examining the meaning to the stage at which the learners formulate the new posed problems in a complete way. At this stage, it is possible to conduct a discussion about each of the questions raised by the learners and examine whether they have a meaning (see below remark regarding the appropriateness of a problem).

Negation of attribute 1: A polygon is given – proposing alternatives:

$(\sim 1)_1$ - an open broken line is given; $(\sim 1)_2$ - a closed broken line intersecting - itself is given; $(\sim 1)_3$ – a circle is given.

A comment:

Alternative $(\sim 1)_2$ is based on the assumption that the learners are familiar with the definition of a polygon as a closed broken line that does not intersect itself (although the mathematical definition of a polygon does not exclude such cases).

Negation of attribute 2: The polygon has three sides - proposing alternatives:

$(\sim 2)_1$ – the polygon has four sides; $(\sim 2)_2$ – the polygon has five sides; $(\sim 2)_3$ – the polygon has *n* sides (or: any number of sides).

Negation of attribute 3: The three sides are equal in length - proposing alternatives:

$(\sim 3)_1$ – only two sides are equal in length; $(\sim 3)_2$ - each of the sides is different in length.

Negation of attribute 4: Calculating the sum of the angles - proposing alternatives:

Please note that this attribute comprises two components, for which alternatives can be proposed: ‘sum’ and ‘angles’.

$(\sim 4)_1$ – calculating the difference between the size of the angles;

$(\sim 4)_2$ – calculating the product of the size of the angles;

$(\sim 4)_3$ – calculating the sum of the length of the heights;

$(\sim 4)_4$ – calculating the sum of the medians length;

$(\sim 4)_5$ – calculating the sum of the angle bisectors length.

Negation of attribute 5: The angles are internal to the triangle - proposing alternatives:

$(\sim 5)_1$ – the angles are external to the triangle.

c. Level 3: Posing new problems based on the proposed alternatives or a combination of several alternatives

At this stage, the learners formulate new problems based on the alternatives proposed at Level 2. The problems can be based on one alternative or on a combination of two or more alternatives. In some cases, the inquiry of the alternatives themselves leads to posing of additional questions and problems, as shown later on.

A comment:

The problems based on alternative $(\sim 3)_1$ – only two sides are equal in length - and on alternative $(\sim 3)_2$ - each of sides is different in length – are solved in a similar way as the given problem (cutting the triangle angles and placing them around a common vertex). Hence, these alternatives are not discussed in this framework.

Note: In all the problems discussed below, it is assumed that we refer to plane geometry.

Problem 1

This problem is based on alternative $(\sim 2)_1$ – the polygon has four sides. The formulation of the new problem is: "An equilateral quadrilateral is given. What is the sum of the equilateral quadrilateral internal angles?".

Please note that in the formulation of the problem, only the number of the polygon sides has changed, whereas the condition of sides equality is maintained.

Solution

The paper of Patkin and Levenberg (Chapter 3 of this book) suggests a solution by cutting the angles and placing them around a common vertex. The problem can also be solved by dividing the quadrilateral into two triangles by means of one of its diagonals (see Figure No. 1). This division allows learners to see that the quadrilaterals have been divided into two triangles. Hence, the sum of the quadrilaterals internal angles is: $180° + 180° = 2 \cdot 180° = 360°$.

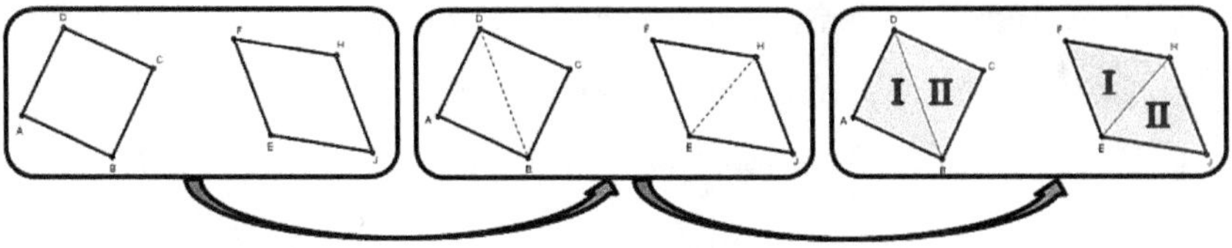

Figure No. 1: Dividing quadrilaterals into two triangles by means of one of their diagonals

Now, it is only natural to integrate alternative $(\sim 2)_1$ with alternative $(\sim 3)_2$: each of the sides is different in length. It is likely to assume that the learners will not find it hard to conclude, even without an additional examination, that in such cases too (and in general, with reference to every quadrilateral), the result will be identical.

Problem 2

This problem is based on alternative $(\sim 2)_2$ – the polygon has five sides.

Since the learners already know that the sum of quadrilateral angles is 360°, regardless of the length of its side, then also in the case of the pentagon, we can relate to a general pentagon.

The formulation of the new problem is: "A pentagon is given. What is the sum of the pentagon internal angles?". Similarly, we can continue asking about polygon with a greater number of sides.

Solution

According to the way of solving problem 2, each of the polygons can be divided into triangles by means of their diagonals (see Figure No. 2). The learners are expected to comprehend that they have to draw only diagonals that come out of one of the vertices, since another division may make it difficult for them to solve the problem.

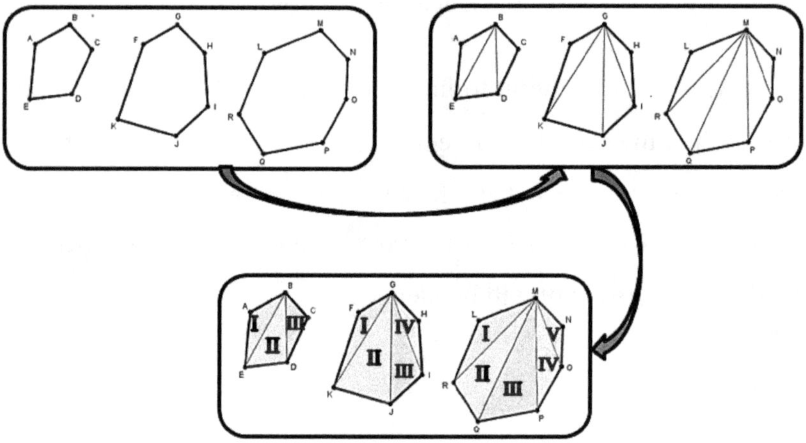

Figure No. 2: Dividing polygons into triangles by means of some of their diagonals

This division enables learners to see that the pentagon can be divided into three triangles, the hexagon into four triangles and the heptagon into five triangles. Consequently, the sum of the polygons internal angles is:

$180^\circ + 180^\circ + 180^\circ = 3 \cdot 180^\circ = 540^\circ$,

$180^\circ + 180^\circ + 180^\circ + 180^\circ = 4 \cdot 180^\circ = 720^\circ$,

$180^\circ + 180^\circ + 180^\circ + 180^\circ + 180^\circ = 5 \cdot 180^\circ = 900^\circ$, respectively.

In this way, the learners will realize that, when a polygon with *n* sides is divided into triangles by means of diagonals that come out of the same vertex, the number of the obtained triangles is *n* - 2. This is because a diagonal is created by connecting two non-adjacent vertices. Each vertex has two adjacent vertices and, therefore, no diagonal can be traced between them. Since *n* - 2 triangles are obtained, then the sum of internal angles in a polygon with *n* sides is $(n-2)\cdot 180^\circ$.

A comment:

- This activity may lead the learners to raise additional questions related to diagonals. These questions could not have been raised following the negation of the given problem attributes since a triangle has no diagonals. This problem is discussed later on (see Problem 5).

Problem 3

This problem is based on the combination of alternative $(\sim 3)_2$: each of the sides is different in length and alternative $(\sim 5)_1$ – the angles are external to the triangle. The formulation of the new problem is as follows: "A polygon with *n* sides is given. What is the sum of the polygon external angles?"

Solution

In the paper of Patkin and Levenberg (Chapter 3 of this book), the learners have to calculate the sum of the external angles by cutting and pasting. Below we suggest a different way. It is based on the findings obtained from problem 2 with reference to the sum of polygon angles, as well as on the learners' previous knowledge that the sum of two adjacent angles is equal to the size of a straight angle, namely 180°.

We start by calculating the sum of a triangle external angles (see Figure No. 3). We have to calculate the sum of $\angle 1 + \angle 2 + \angle 3$. As the figure illustrates, there are three pairs of adjacent angles.

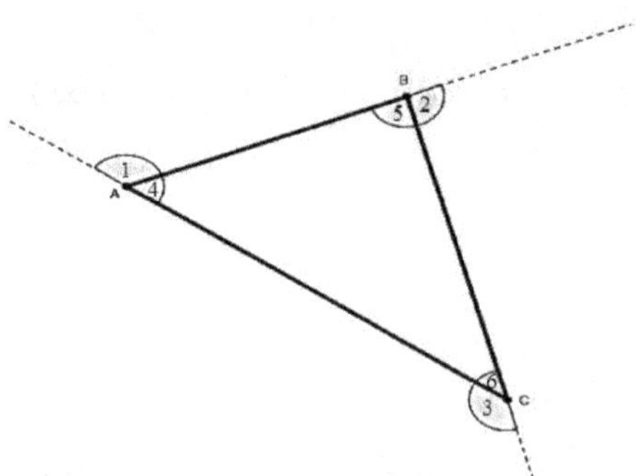

Figure No. 3: External angles of a triangle

Hence, the following values are obtained:

$$(\angle 1 + \angle 4) + (\angle 2 + \angle 5) + (\angle 3 + \angle 6) = 3 \cdot 180° \Rightarrow$$
$$\Rightarrow (\angle 1 + \angle 2 + \angle 3) + (\angle 4 + \angle 5 + \angle 6) = 3 \cdot 180° \Rightarrow$$
$$\Rightarrow (\angle 1 + \angle 2 + \angle 3) + 180° = 3 \cdot 180° \Rightarrow$$
$$\angle 1 + \angle 2 + \angle 3 = 3 \cdot 180° - 180° = 360°$$

That is, the sum of a triangle external angles is 360°.

In a similar way, we calculate the sum of a quadrilateral external angles (see Figure No. 4):

$$(\angle 1 + \angle 5) + (\angle 2 + \angle 6) + (\angle 3 + \angle 7) + (\angle 4 + \angle 8) = 4 \cdot 180^\circ$$

$$(\angle 1 + \angle 2 + \angle 3 + \angle 4) + (\angle 5 + \angle 6 + \angle 7 + \angle 8) = 4 \cdot 180^\circ$$

$$(\angle 1 + \angle 2 + \angle 3 + \angle 4) + 2 \cdot 180^\circ = 4 \cdot 180^\circ \Rightarrow$$

$$(\angle 1 + \angle 2 + \angle 3 + \angle 4) = 4 \cdot 180^\circ - 2 \cdot 180^\circ = 360^\circ$$

And of a pentagon (see Figure No. 5):

$$(\angle 1 + \angle 6) + (\angle 2 + \angle 7) + (\angle 3 + \angle 8) + (\angle 4 + \angle 9) + (\angle 5 + \angle 10) = 5 \cdot 180^\circ$$

$$(\angle 1 + \angle 2 + \angle 3 + \angle 4 + \angle 5) + (\angle 6 + \angle 7 + \angle 8 + \angle 9 + \angle 10) = 5 \cdot 180^\circ$$

$$(\angle 1 + \angle 2 + \angle 3 + \angle 4 + \angle 5) + 3 \cdot 180^\circ = 5 \cdot 180^\circ \Rightarrow$$

$$(\angle 1 + \angle 2 + \angle 3 + \angle 4 + \angle 5) = 5 \cdot 180^\circ - 3 \cdot 180^\circ = 360^\circ$$

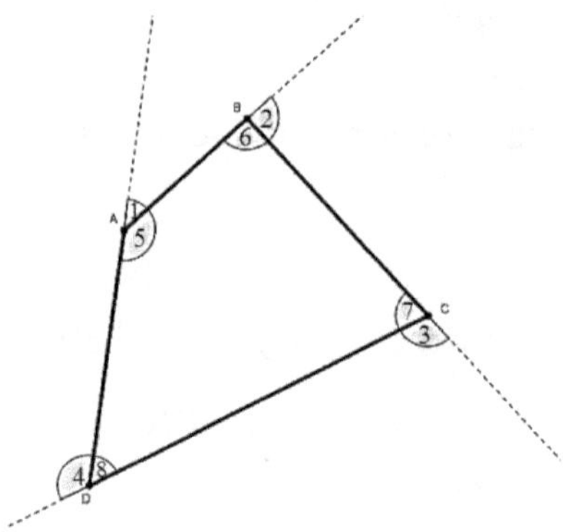

Figure No. 4: External angles of a quadrilateral

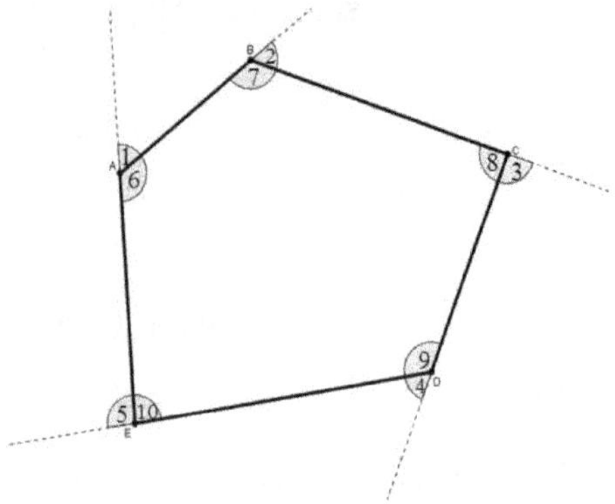

Figure No. 5: External angles of a pentagon

Generally speaking, this examination indicates that for a polygon with *n* sides, *n* pairs of adjacent angles will be obtained by the process described above. The sum of angles of these *n* pairs is $n \cdot 180^\circ$. In order to calculate the sum of the external angles, the sum of the polygon internal angles should be subtracted from $n \cdot 180^\circ$, namely $(n-2) \cdot 180^\circ$. Hence, the obtained result is:

$$n \cdot 180^\circ - (n-2) \cdot 180^\circ = n \cdot 180^\circ - (n \cdot 180^\circ - 2 \cdot 180^\circ) = 2 \cdot 180^\circ = 360^\circ$$

The total sum of the external angles of any polygon is 360°. Given that the sum of a polygon internal angles depends on the number of its sides, the discovery concerning the sum of the external angles is undoubtedly surprising.

Problem 4

This problem is based on the combination of alternative $(\sim 1)_2$ – a closed broken line intersecting itself is given and alternative $(\sim 2)_3$ – the polygon has n sides (or: any number of sides). As mentioned above, from a geometric point of view, a polygon is defined as 'a closed broken line', without adding the condition 'that does not intersect itself', although at school it is customary to add this reservation.

The new problem, then, is formulated in the following way: "A shape, that is formed by a closed broken line that interests itself and that consists of n sides, is given. What is the sum of the internal angles of this shape?".

Solution

We will answer this question in a gradual manner. We start by searching for the solution of the problem: "A shape that is formed by a closed broken line that interests itself and that consists of four sides, is given. What is the sum of the internal angles of this shape?".

Figure No. 6 shows a shape with four sides. In fact, from a geometric point of view, this concerns a quadrilateral, even though in the school framework it is not customary to relate to this shape as a quadrilateral. This shape is a quadrilateral because, like any other quadrilateral, it has four sides – *AB, BC, CD* and *DA* and four angles (indicated in the figure). The diagonals of this quadrilateral – *AC* and *AD*, are external to the quadrilateral (diagonals that are outside the polygon are known to us from concave polygons).

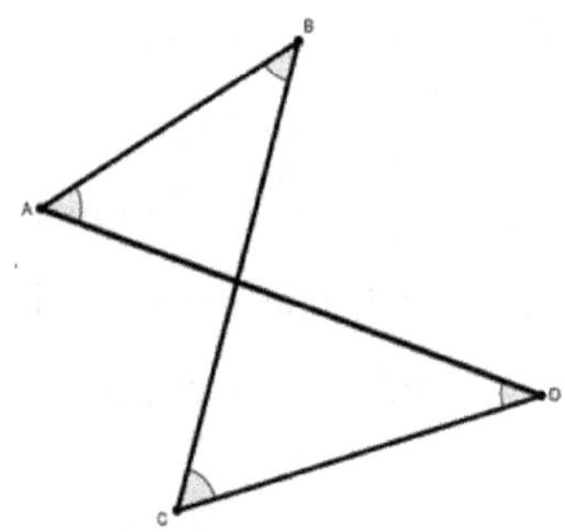

Figure No. 6: A quadrilateral made of a broken line that interests itself

It is easy to see that the sum of the internal angles of such a quadrilateral is not fixed. It is smaller than 360° since the sum of its four angles is smaller than the sum of the angles of two triangles. This is apparently why such a quadrilateral is not discussed in geometry lessons. However, teachers can conduct in class an interesting debate on the question: “Is this shape a quadrilateral?”

We examine now whether the same applies to a closed broken line that interests itself and consists of five sides (see Figure No. 7). The obtained shape is called ‘a star’ and we have to calculate the sum of the five angles indicated in the figure, namely the sum of

$$\angle 1 + \angle 2 + \angle 3 + \angle 4 + \angle 5$$

In order to calculate the sum of these angles, we apply our previous knowledge concerning the sum of external angles in a “common” polygon (one that consists of a line that does not intersect itself), based on the solution of Problem 3. In addition, we apply the theorem "an external angle in a triangle is equal to the sum of the two angles that are not adjacent to it". Learners who are not familiar with this theorem can discover it easily by making clippings of angles.

As Figure No. 8 illustrates, the five-pointed star is enclosed by a convex pentagon *ADBEC* and delimits a convex pentagon *FGHIJ*. As we saw in problem 2, the sum of the angles of the convex pentagon *ADBEC* *is* $3 \cdot 180^\circ = 540^\circ$, and as we saw in problem 3, the sum of the external angles of the convex pentagon *FGHIJ* is 360°.

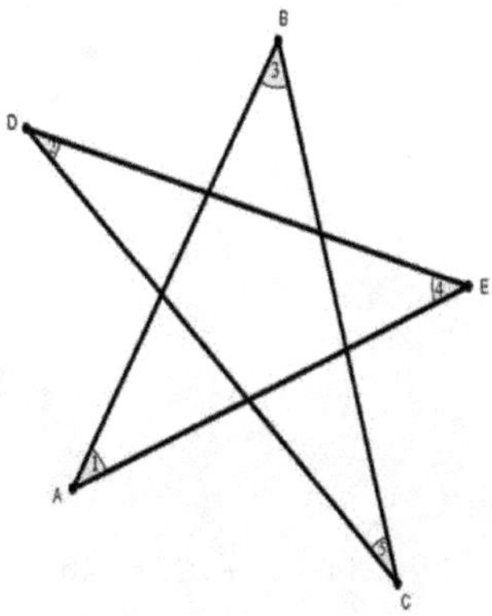

Figure No. 7: A five-pointed star

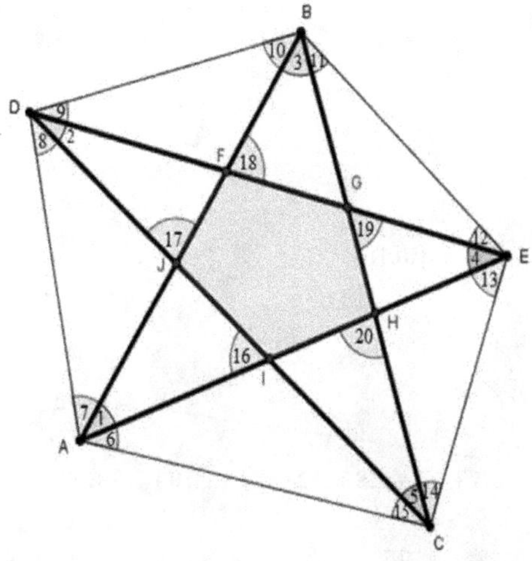

Figure No. 8: A five-pointed star enclosed by and delimits a convex pentagon

Hence, the obtained result is:

$$\angle 6+\angle 1+\angle 7+\angle 8+\angle 2+\angle 9+\angle 10+\angle 3+\angle 11+\angle 12+$$
$$+\angle 4+\angle 13+\angle 14+\angle 5+\angle 15=540^\circ$$
$$\angle 16+\angle 17+\angle 18+\angle 19+\angle 20=360^\circ$$

Now, we are going to apply the theorem engaging in a triangle external angle. For example, $\angle 16$ is an external angle of triangle *AIC*, hence $\angle 16=\angle 6+\angle 15$. Similarly,

$$\angle 17=\angle 7+\angle 8$$
$$\angle 18=\angle 9+\angle 10$$
$$\angle 19=\angle 11+\angle 12$$
$$\angle 20=\angle 13+\angle 14$$

Consequently,

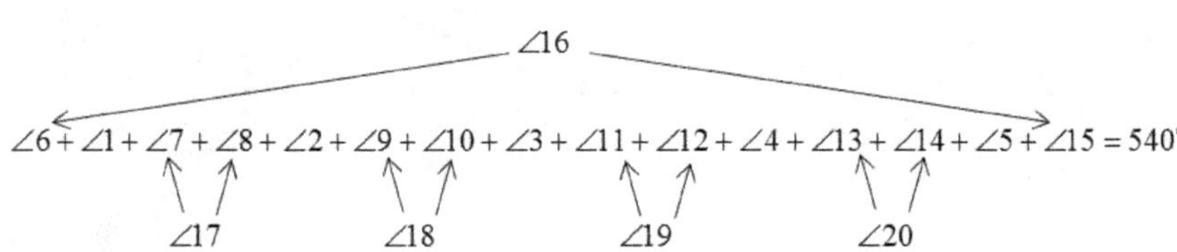

And thus

$$\angle 6 + \angle 1 + \angle 7 + \angle 8 + \angle 2 + \angle 9 + \angle 10 + \angle 3 + \angle 11 + \angle 12 +$$
$$+\angle 4 + \angle 13 + \angle 14 + \angle 5 + \angle 15 = 540°$$

$$(\angle 1 + \angle 2 + \angle 3 + \angle 4 + \angle 5) + (\angle 16 + \angle 17 + \angle 18 + \angle 19 + \angle 20) = 540°$$

$$(\angle 1 + \angle 2 + \angle 3 + \angle 4 + \angle 5) + 360° = 540°$$

$$\Rightarrow \angle 1 + \angle 2 + \angle 3 + \angle 4 + \angle 5 = 180°$$

The result is that the sum of the internal angles of a five-pointed star is 180°.

The readers are encouraged to continue exploring the sum of internal angles of stars. Through the activity, they can check whether stars with an even number of sides can be formed so that each of the star vertices is part of a triangle. Please note that the shape described in Figure No. 9 is not a star but the shape described in Figure No. 10 is, in fact, a star.

Furthermore, the readers are invited to discover the generalization of the case of a five-pointed star: the sum of the angles of a star with n sides, when n is an odd number, is: $(n - 2) \cdot 180° - 360°$. As seen in the case of the five-pointed star, this sum is obtained in fact, by calculating the sum of the internal angles of a convex polygon with n sides and, then, subtracting the sum of the polygon external angles.

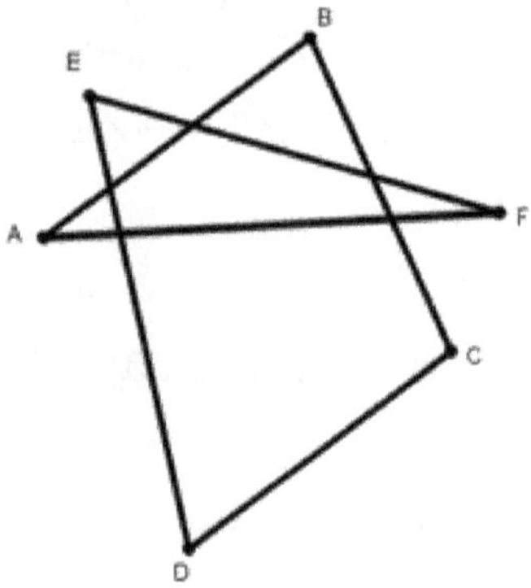

Figure No. 9: A non-star shape

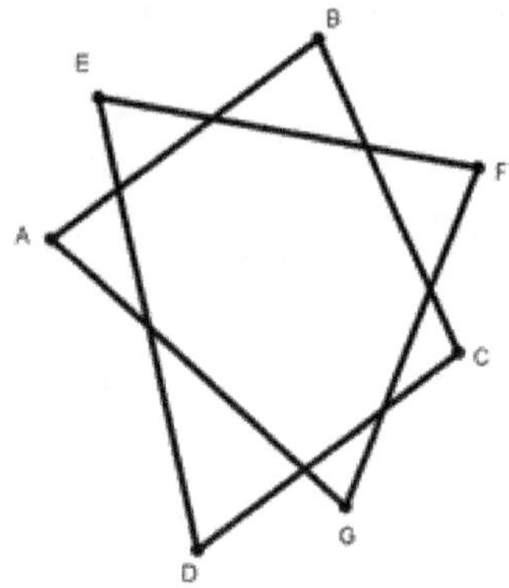

Figure No. 10: A star shape

Problem 5

In problem 2 we used the diagonals of a polygon in order to find out the number of triangles formed as a result of dividing the polygon as described in Figure No. 1 and Figure No. 2. This activity could lead the learners to raise additional questions related to polygon diagonals. For example: "What is the number of diagonals in a convex polygon with n sides?".

Solution

Let us look at Figure No. 11. The quadrilateral has two diagonals, the pentagon has five diagonals and the hexagon has 9 diagonals. In order to count the diagonals, it is recommended to coloring every diagonal that has already been counted. What is the relation between the number of the convex polygon sides and the number of its diagonals?

As indicated above, a diagonal connects two vertices that are not adjacent to it. Every vertex has two adjacent vertices and it is clear that a diagonal that connects a vertex to itself cannot be traced. Consequently, in a polygon with n sides, $n-3$ diagonals pass through each vertex. Since the polygon has n sides (or n vertices), we should apparently have obtained $n\cdot(n-3)$ diagonals. However, please note that such a calculation is feasible only if we count each diagonal twice, since, for example in the case of the quadrilateral presented in Figure No. 11, the diagonal *BD* is exactly the diagonal *DB*. Hence, the number of diagonals in a convex polygon with n sides is $\frac{n\cdot(n-3)}{2}$.

Please check that this formula is indeed valid for the three polygons presented in Figure No. 11.

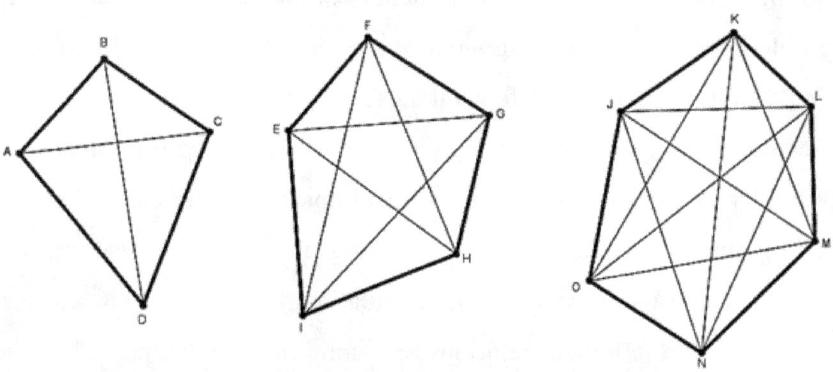

Figure No. 11: Polygons and their diagonals

Following this problem, learners can pose additional problems related to polygon diagonals. For example, a problem that concerns the number of polygons created by these diagonals and their type thereof. We will not expand on this issue within the scope of this chapter.

Summary

This chapter illustrates how we can start with a simple given problem and, by applying the WIN strategy, advance to complex problems. Sometimes, new problems emerge from problems posed on the basis of the given problem, as demonstrated by problem 5. This strategy allows teachers to easily turn every problem from the textbooks into an inquiry problem that leads the learners to an in-depth discovery. Thus, it facilitates the development of their inquiry skills and, at the same time, develops their mathematical creativity.

Furthermore, according to Haylock (1986), every good mathematical assignment should comprise a component of problem posing, a component of problem solving and a component of a re-definition. As could be seen, the WIN strategy includes, in fact, all these three components. At the Level 1 – creating a list of the given problem attributes – learners have the opportunity to re-think about mathematical objects and terms. At Level 2 – raising the WIN question and proposing alternatives – learners are required to re-define the mathematical situation. In order to do it, learners have to refer to the logical relations and conditions encompassed in the problem, as well as determine how to gather any data for creating a coherent mathematical situation (Lavy & Shriki, 2010). Level 3 – posing new problems and searching for a solution for one or more of them – is obviously associated with problem solving.

Finally, readers are encouraged to apply the WIN strategy to geometric problems that appear in the other chapters of this book.

References

1. Brown, S. I., & Walter, M. I. (1969). What if not? *Mathematics Teaching, 46,* 38–45.
2. Brown, S. I., & Walter, M. I. (1990). *The art of problem posing*. Hillsdale, NJ: L. Erlbaum Associates.
3. Cunningham, R. (2004). Problem posing: An opportunity for increasing student responsibility. *Mathematics and Computer Education, 38*(1), 83–89.
4. English, L. D. (1997). Promoting a problem-posing classroom. *Teaching Children Mathematics, 4,* 172–179.
5. Ellerton, N. F. (1986a). Children's made-up mathematical problems: A new perspective on talented mathematicians. *Educational Studies in Mathematics, 17,* 261–271.
6. Ellerton, N. F. (1986b). Mathematics problems written by children. *Research in Mathematics Education in Australia, December,* 32–44.
7. Haylock, D. W. (1986). Mathematical creativity in schoolchildren. *Journal of Creative Behavior, 21*(1), 48–59.
8. Lavy, I., & Shriki, A. (2010). Engaging in problem-posing activities in a dynamic geometry setting and the development of prospective teachers' mathematical knowledge. *Journal of Mathematical Behavior, 29,* 11–24.
9. Levav-Waynberg, A., & Leikin, R. (2012). Using multiple solution tasks for the evaluation of students' problem-solving performance in geometry. *Canadian Journal of Science, Mathematics and Technology Education, 12*(4), 313–333.
10. Nikata, K., & Shimada, H. (2005). Facilitation of analogical transfer by posing an analogous problem for oneself. *Japanese Journal of Educational Psychology, 53,* 381–392.
11. Shriki, A. (2004). Being a real mathematician – Everyone can do it! *Alex – Bulletin for Mathematics Teachers, 31*, 4–13. [Hebrew].
12. Shriki, A. (2007). When you investigate – you discover. *Alex – Bulletin for Mathematics Teachers, 37*, 101–106. [Hebrew].
13. Shriki, A. (2013). A model for assessing the development of students' creativity in the context of problem posing. *Creative Education, 4*(7), 430–439.

14. Shriki, A., & Lavy, I. (2012). Problem posing in a dynamic geometry environment and the development of mathematical insights. *The International Journal of Learning, 18*(5), 61–70.
15. Silver, E. A. (1997). Fostering creativity through instruction rich in mathematical problem solving and problem posing. *ZDM - The International Journal on Mathematics Education, 29*(3), 75–80.
16. Silver, E. A., & Cai, J. (1996). An analysis of arithmetic problem posing by middle school students. *Journal for Research in Mathematics Education, 27,* 521–539.
17. Stoyanova, E., & Ellerton, N. F. (1996). A framework for research into students' problem posing. In: P. Clarkson (Ed.), *Technology in mathematics education* (518–525). Melbourne: Mathematics Education Research Group of Australasia.

Section III

Geometry and the Creative Brain

Photo by S. Gazit

https://doi.org/10.1142/9789811218750_0008

Chapter 8

From Peeling a Tangerine to Image Processing: The Beauty of Geometry and Its Relevance to Everyday Life

Eli Appleboim

Introduction

Geometry helps doctors to save lives.

This chapter aims to infuse content into this statement, as prestigious as it sounds, convincing the readers that this does not concern a meaningless phrase but rather a totally realistic description.

Geometry has been one of the basic areas of mathematics since the dawn of history until our present days, both as a research field and a major topic of mathematics teaching at school. Throughout the years, the charm and appeal of geometry have not faded in the eyes of mathematicians. Today, it still constitutes a central research field that is constantly developing, branching out to a wide variety of topics, both in mathematics and in other sciences. Conversely, learners quite frequently perceive geometry as a field that is far from fascinating but is also frightening and rebuffed.

One of the problems associated with the unappealing image of geometry is its consideration (erroneous of course) as an outdated and stagnant subject. Geometry as a research field transcended already 200 years ago the boundaries of the plane and solid geometry, whose rules and mainly the system of axioms defining it had been determined by the Greek mathematician Euclid more than two thousand years ago. However, within the framework of geometry studies at school, learners are still exposed to the same ancient system of axioms. Exposure to more updated aspects of geometry, particularly to models of non-Euclidean geometry, is shared by few, if any, learners, mostly those excelling in mathematics studies.

Hence, it is incumbent on us to find ways of drawing geometry closer to the heart of as many learners as possible. Moreover, we must demonstrate to them its beauty, importance, and relevance in the present age. This chapter suggests such a layout by teaching the topic of image processing.

Unlike ancient geometry, image processing is a new area that has developed in the modern age. This is a research area of applied mathematics, whereby computerized mathematical instruments are applied in order to understand the content, analysis, and enhancement of various types of images (photos taken with a regular home camera and under normal conditions, medical images, radar images, and others). This research area was conceived in fact following the digital revolution of the 1950s. From its inception, this area has been grounded in mathematical elements that are much more modern than geometry, mainly in harmonic analysis and Fourier theory that were developed in the 19th century.

Can geometry contribute to the innovative and rapidly developing area of image processing? This chapter attempts to show that the answer to this question is definitely positive. Starting from the 1990s and even more so since the mid-2010s, geometry and mainly non-Euclidean geometry, has become a most central tier in image processing as well as in other topics from the fields of engineering and computer sciences, such as computerized graphics and artificial intelligence.

As a function of the numerous examples presented in this chapter, I hope to transmit the message regarding the importance of expanding the study of geometry and inculcating terms from non-Euclidean geometry already at the learning stages at school. The objective is to prepare learners for integration at the head of research areas and technological industry, whereby mathematics in general and geometry, in particular, occupy an increasingly growing central place. Moreover, this would inculcate original thinking competencies and a wider picture of the world in the entire learner population (Henderson, 2005). At the end of the chapter, I invite the readers to share my personal experience in this field and my perception of geometry teaching.

The first part of this chapter is dedicated to a short theoretical background of non-Euclidean geometry. A key term defined in this part is 'curvature'. This is followed by a concise review of image processing. The second part presents three of the main problems of image processing with which researchers are coping. The third part discusses solutions to these problems. After a brief discussion on classical solutions, most of the chapter will focus on the geometric approach to image processing. In this context, I will present geometry-based solutions to the problems presented in the second part. Finally, the fourth part focuses on the teaching of geometry to learners at different ages and different levels. Within this framework, I will suggest examples for demonstrating key terms, problems, and solutions. The chapter ends with a description of my personal experience.

Theoretical Background

Development of non-Euclidean geometry

About the year 1820, the German mathematician, **Carl Friedrich Gauss** (1777–1855) (Figure 1) performed a survey of geographic measurements around Germany.

Figure 1: Carl Friedrich Gauss

Throughout the survey, Gauss noticed that his measurements did not correspond to his earlier calculations. For example, he turned in a circle around the center of Munich, constantly maintaining a distance of 200 km from the center of the city. He found out that the perimeter of the circle along which he turned was **shorter** than the perimeter calculated according to the formula learned at school, a formula that indicated the relationship between the perimeter of a circle and its radius:

Perimeter of a circle = (radius of the circle * 2π)

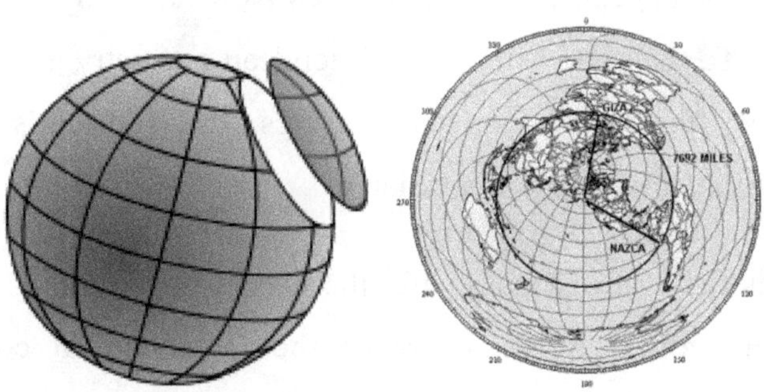

Figure 2: A spherical circle around some point - a view from the side and a view from above

Gauss realized that the fact that the measurement results were not in line with the above-mentioned geometric rule, was tightly connected to the fact that Planet Earth has a spherical shape. Conversely, the rule specifying that the perimeter of the circle depends on its radius applies to circles drawn on a planar surface, such as the class blackboard or a desk. This insight evoked his interest in the geometry of Planet Earth and, more generally, in geometry of surfaces that are not planar. From now on we refer to these surfaces as **curved surfaces**. Gauss understood that when we try describing the geometry of these surfaces, Euclid's axiomatic system that defines Euclidean geometry is no longer valid and a new axiomatic system should be defined. This gave rise to the first thoughts about non-Euclidean geometry. Gauss did not publish these discoveries and ideas, for fear of incurring the wrath of the church. The reason was that for many hundreds of years, it was assumed the Euclidean geometry described nature and its laws that had been determined by God Almighty. Hence, undermining the rules of geometry might have been considered tantamount to undermining the basic laws of nature. Today we know for sure that Euclidean geometry describes well Newton's laws of mechanics for example. To the same extent, non-Euclidean geometry describes with great accuracy the laws of modern physics, like Einstein's theory of general relativity and quantum mechanics.

Towards the end of Gauss' life, the need for expanding geometric concepts was voiced by members of the mathematical community. Moreover, several non-Euclidean geometry models were proposed. However, nothing was yet formulated in a sufficiently explicit way. All this was changed by **Bernard Riemann** (1826–1866) (Figure 3). Riemann, Gauss' most gifted disciple, submitted to his master his doctoral thesis that engaged in the basic axioms of geometry and the issue of their validity in general. Riemann was the first to try determining accurate characteristics of the validity of various geometric rules. His thesis was the first fluent and coherent mathematical treatise that dealt with issues, such as the relation between a certain surface being curved and the validity of the various theorems, rules, and formulas of Euclidean geometry and the way by which they have to be changed. For example, the question of the validity of the rule according to which in order to calculate the perimeter of a circle, the circle radius is multiplied by twice the constant size π. Another example that seems even more far-reaching, is the validity of Euclid's axiom according to which "between two points only one straight line can be drawn". This axiom is not valid for some of the curved surfaces.

Figure 3: Bernard Riemann

Riemann and his geometrician disciples showed that the most basic geometric laws crucially depended on the extent to which the surface one attempted to describe, was "far from" being planar. The basic term that quantifies this size is called **curvature**. Curvature is a property of geometric shapes and it plays a key role in the field of image processing. The definition of the curvature is elaborated below.

Curvature and its geometric meaning

The definition of curvature mainly aims to quantify to what extent any surface is not planar. This quantification directly affects various geometric rules, such as the validity of the axiom "between two points only one straight line can be drawn".

Gauss encountered a problem in his measurements, namely that they were not in line with the rule stipulating that the perimeter of a circle is equal to the multiplication of the radius by 2π. This problem leads us in fact to the following formula. Let's assume that C is a circle around a given point p on a given surface. We indicate the radius of the circle C with the letter R and its perimeter by the letter L. Hence,

$$\text{Curvature} = \frac{2\pi R - L}{R}$$

It is noteworthy that the curvature formula illustrates that the obtained size depends on the radius of the circle we decide to draw. If we chose another circle, with another radius, we might have obtained a different size for the curvature. In this respect, the formula is only partial. To use the mathematicians' language, the curvature has not been well defined. In order to overcome this difficulty and avoid being dependent on a specific circle, the limit basics of differential calculus is applied. Every time we choose another circle with a gradually decreasing radius until finally, the curvature is the size obtained when the radius of the circle approaches zero (Do Carmo, 1976; Spivak, 1999).

The following are several main properties of the curvature of surfaces (Do Carmo, 1976; Henderson, 2005; Spivak, 1999):

1. The curvature of the plane is 0. This fact stems directly from the definition since the perimeter of a plane circle with radius R is exactly $2\pi R$.

2. The curvature of a plane does not change if we shift or rotate it in space.

3. The curvature of a convex surface, namely a surface shaped like an egg, or an ellipse that is not entirely symmetrical, does indeed change from point to point but at every point it is positive. And vice versa: if the curvature of a plane is positive at any place, then the plane is convex. Such a surface is called an **elliptic surface** and the geometry of such surfaces is **elliptic geometry** or **spherical geometry**. Planet Earth is an elliptic surface (see Figure 4).

Figure 4: An elliptic surface

4. The curvature of a plane that looks like the saddle in Figure 5 is negative in every place.

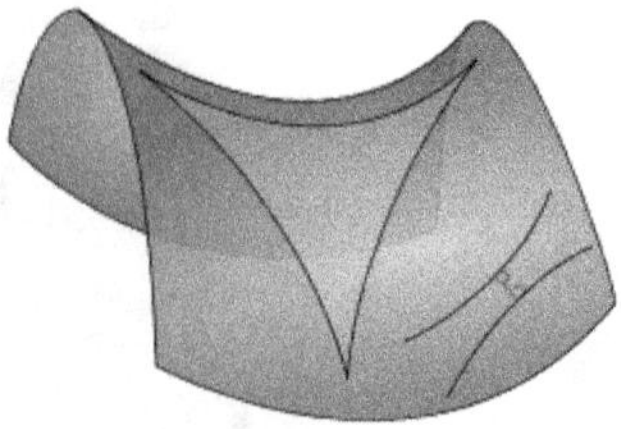

Figure 5: A surface with a negative curvature

Such a surface is called a **hyperbolic surface** and its geometry is **hyperbolic geometry**. In this case, the opposite is also true: if the curvature at any point on a surface is negative, the surface looks like a saddle at every point and is hyperbolic. A further example of such a surface is demonstrated in Figure 6.

Figure 6: A hyperbolic surface

5. In spherical geometry and hyperbolic geometry, Euclid's parallel postulate does not apply. This axiom stipulates that one and an only straight line can be drawn parallel to another given straight line through an external point.

6. Not only is the curvature in any shape not constant but it can be negative, positive or 0 in different zones. Figure 7 demonstrates zones whose curvature is positive, zones whose curvature is negative and others whose curvature is 0.

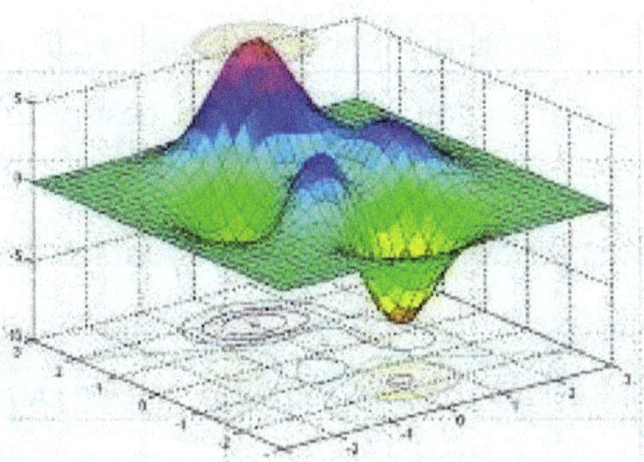

Figure 7: A general surface with a changing curvature

7. In case that a surface is curved, i.e. its curvature is other than 0, even the straight lines on such a surface should be re-defined. Consequently, we usually do not relate to the straight line between two given points on a curved surface but to the shortest line between the two points.

This part of the chapter that discusses curvature ends with a comment, namely that curvature can be defined also for a higher dimension, for example, solids with volume. This is a most crucial fact in the theoretical mathematical study of geometry and other fields as well as in applied research. Thus, for example, the curvature properties of the 3-dimensional universe in which we live, enable us to scientifically prove the existence of black holes in space (Yau, 2001).

A short introduction to image processing

The digital era is tightly linked to the development of image processing. The fact that a digital image is represented by a collection of numbers that can be easily read by a computer, facilitates supplying the computer with capabilities of "seeing" and "understanding" the world around it. Image processing is the scientific field, whereby advanced methods of image changing are developed so that eventually the best results are attained by pressing a button. It is noteworthy that, as yet, there are no accurate mathematical indices for determining when an image is good and when it is not. Finding objective indices is one of the major unsolved problems of research in this field. Nevertheless, images can be defined as having a contract level that is better than others, sharper than others or smoother than others. These indices can be accurately calculated and they sometimes serve as quality indices.

When thinking about the term "image", we usually think about pictures photographed by a standard camera (Figure 8). However, in fact, there are varied means of creating images, such as medical imaging machines, cameras for pictures with a depth dimension, radar cameras, and others. Image processing does concern a much wider variety of images, including medical ones (ultrasound, CT, MRI), images with a depth dimension that contain 3-dimensional information, etc.

Figure 8: A film camera and a typical picture

The research and applications related to image processing have developed in parallel to the development of photography means throughout the years. The significant breakthrough of this field as a research topic in academia and as a main development field in the industry transpired with the conception of digital photography. Apparently, there is no difference between a digital camera and a film camera of previous generations. Both cameras consist of a metal pack with a lens on one side. Light penetrates through the lens, is projected on a board or on a film coated with a light-sensitive material which is on the other side of the camera. Moreover, the nature of the light-sensitive material in a digital camera is identical to that in the film camera but the way of turning it into a picture is entirely different. In a film camera, the light projected on the film spreads all over the film in a continuous way. Then, a picture is produced by a development process that involves the use of chemical substances. Conversely, in a digital camera, the light is absorbed by a set of cells, each containing the light-sensitive material. Each such cell is called a **picture element** or by its more common name **pixel** (= picture element). The process can be described in the following manner: instead of producing a picture in a chemical process, the picture is produced by an electric process whereby the light absorbed by a receptors board, is translated into a number layout. The number of pixels on the receptors board represents the resolution of the picture. The more pixels there are, the better the distinction of details in the picture is. This is demonstrated by Lena's picture in Figure 9. Lena is a Swedish model whose picture in one of Playboy issues of the 1970s serves as a test picture for many algorithms of image processing. In addition to the amusing anecdote about the source of the picture, it is considered as a good benchmark picture since it contains many details, e.g. the hat feathers, eye pupils and so on. Many of the image processing algorithms are tested among others by their ability to maintain the sharpness of these details. Figure 9 below presents two pictures of Lena, one at high resolution and the other at low resolution.

In the picture on the left, there are many more pixels and hence every detail in the picture has a pixel of its own and does not compete with other details for the same pixel. Therefore, this picture is sharper and clearer.

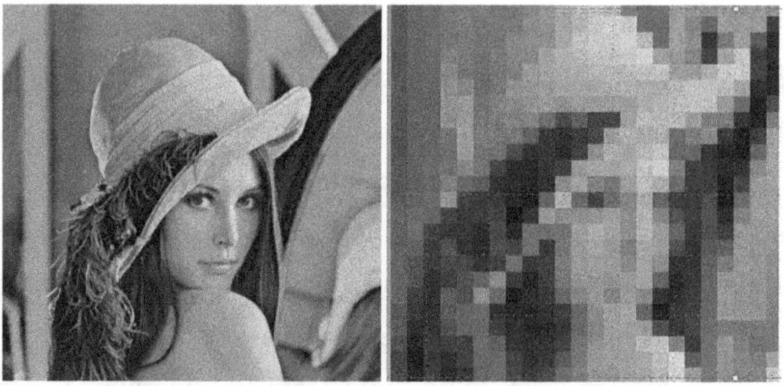

Figure 9: On the left, Lena's picture at high resolution and on the right her picture at low resolution

Prior to the digital photography age, not much could be done in order to affect the quality of the picture after it was taken. Once the picture was absorbed on the film, its quality could be enhanced only by very limited chemical means. Conversely, since a digital picture is a collection of numbers stored in the memory, each of the numbers can be accessed and changed, changing the picture accordingly.

Main Problems in the Field of Image Processing

The problems that researchers in the field of image processing cope with are numerous and varied and all of them cannot be included within the scope of this chapter. Below are three of the most crucial problems that have preoccupied the researchers in this field. Almost every new method or algorithm that is suggested attempts to cope at least with one of these problems (the order of their presentation is random and insignificant):

Image denoising

Figure 10 shows Lena's picture but it is obvious that the picture is not on high quality as that in Figure 9. The difference in quality is due to the fact that the current picture seems "dirty" and significantly less sharp. Lena's face and feathers of her hat look full of "pimples".

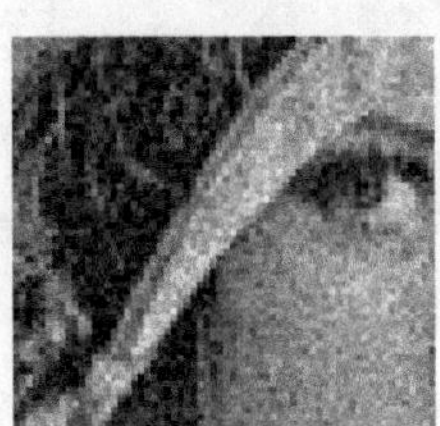

Figure 10: A picture full of noise

The accepted term for such faults in the picture is **noise**. It is borrowed from the field of audio signals processing. It is similar to the occurrence of background noises heard when listening to the radio when the device is not tuned to the exact frequency of the station to which one listens. Imagine that you are trying to eliminate these background noises but without moving the radio dial to the right frequency of the station. The mathematical challenge you are facing is finding (or not finding) a mathematical function or a series of mathematical functions that will remove the noise as if you have moved the radio dial to the right frequency. One of the amazing wonders of mathematics is that one can find or invent such functions.

Sharpening blurry areas

Figure 11 shows once more Lena's picture. However, this time, the damaged quality of the picture is manifested by the fact that it seems blurred as if taken "out of focus". The challenge, in this case, is finding a mathematical solution that will sharpen the picture so that it looks as if we have taken it once more, this time with precise tuning of the lens focus.

Figure 11: A blurred picture was taken out of focus

Segmentation

The problem of dividing a picture into different segments according to their content (e.g. healthy tissue versus a suspicious one, or identification of street lights in a picture taken by a camera installed in a moving vehicle) is a problem of segmentation. Figure 12 demonstrates a medical image in which the outlines of an area are highlighted. This area is suspected as a cancerous tissue. The medical challenge, in this case, is obviously identifying this suspicious tissue.

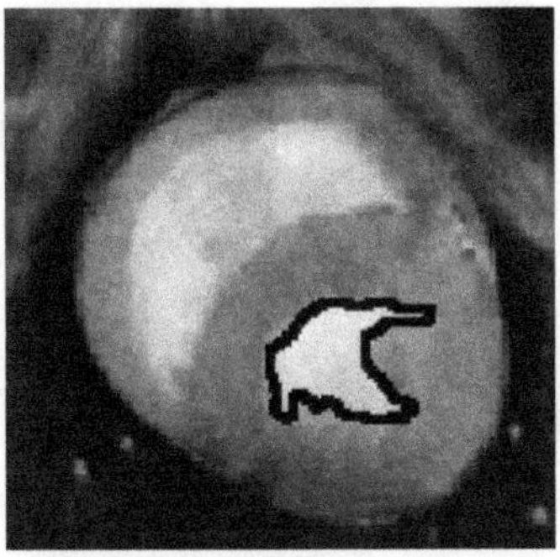

Figure 12: Marking of a suspicious tissue in a medical image

In the image demonstrated in Figure 12, the challenge does not seem difficult since the suspicious area is big and quite obvious. The medical challenge would have become much more complicated if the clear spot marked in the image were very small, just a few pixels and if its level of clarity were more similar to the background around it. One should remember that in many cases, doctors are required to go over a great number of such images. A CT scan of the colon, for example, yields about 1000 images and in the case of a chest scan, doctors have to check about 700 images. Going over such a high number of images, some of which comprise very little suspicious areas that are not much different than their environment, is a very exhausting task. Performing this task while paying maximal attention to the most minute details and at an extremely high level of accuracy, constitutes a much harder medical challenge. The image that those hundreds of images could be put into a machine (a mathematical machine in this case) that would have found for the doctor these suspicious areas in an automatic, relatively rapid and highly accurate manner. All the doctors would have to do is to quickly check the findings demonstrated by the machine in order to corroborate them. In this case, the doctor's life would have been much easier with an incomparable smaller number of diagnosis errors. Building such a machine turns the medical challenge into a mathematical challenge.

Solutions of the Problems Presented in Part 2

Classic solutions

The classic solutions to the problems presented above and to other problems are based on the use of filters. The Filter theory is a regular mathematical theory from the field of harmonic analysis. Thus, for example, for the purpose of image denoising, we use a filter that changes the value of every pixel to the mean values of the pixels around it. The size of the filter determines the size of the environment in which the mean value is calculated. This process results in a new picture in which the pixel values are closer to each other. Consequently, the picture looks more uniform and less noisy (Gonzalez & Woods, 2008).

New direction – Geometric approach to image processing

Although classic filters manage to cope with the various challenges of image processing, their results are far from perfect. For example, the filter used for image denoising reduces the degree of contrast of the picture and, hence, causes an undesirable blurring. This entailed the need for improving the classic methods and thinking about additional instruments. One of the people who engaged in this matter was the American mathematician David Mumford. Mumford, a geometrician laureate of the Fields Medal for his studies in algebraic geometry, experienced the need for improving the image processing instruments in 1985. Due to the limitations of the classic methods prevalent then, the malignant tumor from which his wife suffered was not properly diagnosed. Following her death, Mumford began exploring the question: "Can and in what way the good old geometry, or perhaps the modern one, help to improve the instruments in the field of image processing?"

Mumford conceived the idea of turning in the direction of geometric thinking. It was based on the fact that since an image is actually a function of the longitude line and the altitude line that describe the location of every pixel and since a function of two variables describes a surface, one can easily obtain a representation of an image as a surface (see Figure 13). Such a representation enables reference to the geometric characteristics of the surface, examining their effect on the quality and content of the obtained image (Mumford & Shah, 1985).

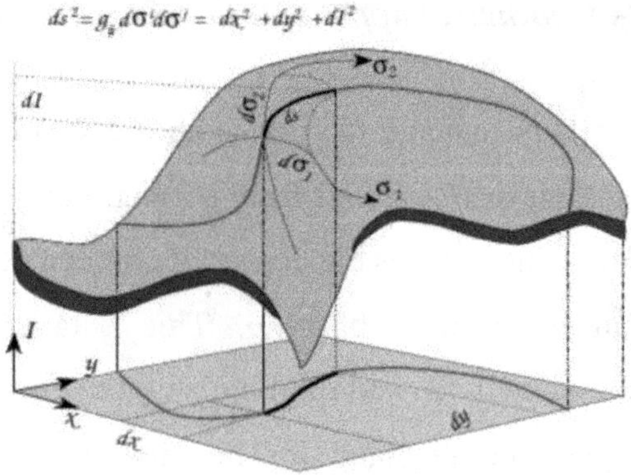

Figure 13: Representation of an image as a surface

Hence, the geometric property of the curvature described before in the purely mathematical context is crucially important in different and varied processes of image processing. High curvature (in absolute value) for example characterizes points at which the picture is noisy. An intermediate curvature is typical of points through which outlines of different objects in the pictures pass. Low curvature usually indicates homogeneous areas (in which the pixels have similar illumination values) in the picture. Consequently, developing filters and filtering methods that take into consideration characteristics of curvature of every point, might overcome the main challenges of image processing, without introducing undesirable disturbances typical of classic filters.

The next sections describe the way by which geometry can be used for improving the quality of images and attaining optimal results while focusing on the three problems presented above.

Image denoising

Figure 14 on the left shows an original picture that contains a noise, in the middle, the same picture is presented after being filtered with a regular filter as described above and, on the right, we see the picture after filtering with a geometric filter, in which the reaction of every pixel is a function of the curvature size in the pixel. One can clearly notice the difference between the results achieved by the two different filters.

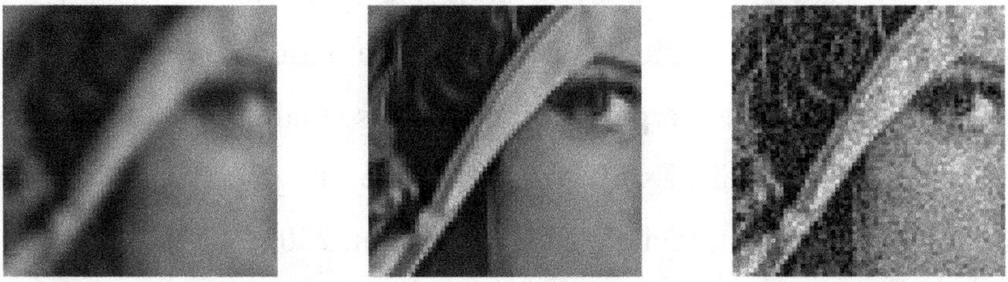

Figure 14: Image denoising

Figure 15 demonstrates the process more clearly. The surface on the left represents part of the noisy picture in Figure 14. It is easy to distinguish on the surface the jagged points of noise whose values of curvature are particularly high. On the right, we see the surface after the geometric filtering. It is obvious that the points of noise have been flattened whereas the main step that represents an outline in any picture has remained unchanged. Thus, for example, the eye outlines remained with high contrast, and so they should be (Sochen & Zeevi, 1988; Spira, Sochen & Kimmel, 2005).

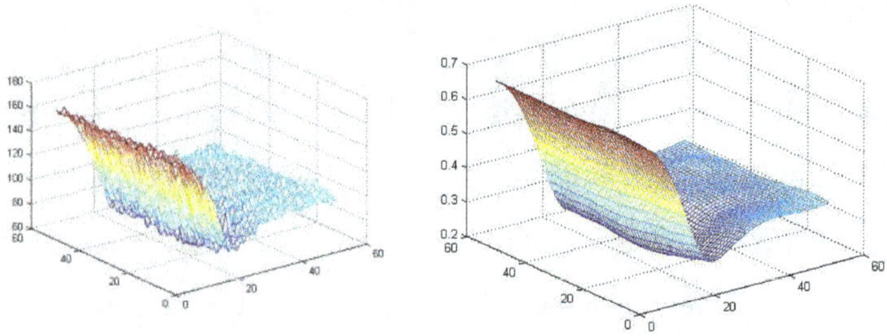

Figure 15: Above is the noisy picture and below – the picture after geometric filtering

Filtering noises while maintaining the sharpness of the picture is extremely important in medical imaging. The reason is that filtering that does not maintain the sharpness might blur essential findings and render it difficult for the medical staff to analyze the image correctly, fully and accurately (Angenent, Pichon, & Tannenbaum, 2006).

Figure 16a and 16b demonstrate the improved result of image denoising of an MRI image in the geometric methods versus other filtering methods. The outlines following the geometric filtering are sharper and clearer than those after filtering with a regular filter and, thus, they facilitate the analysis of the image. On the left side of Figure 16b, we see distinctly the sharpness that is typical of a geometrically-filtered image.

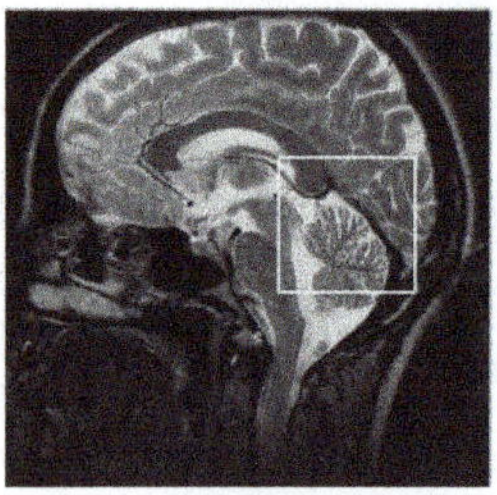

Figure 16a: An MRI image of the brain

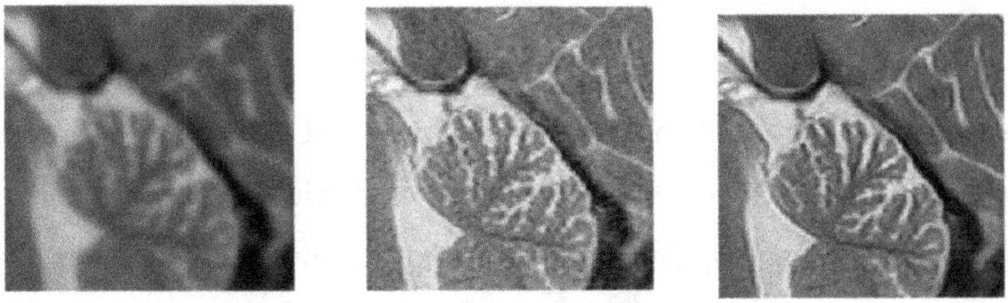

Figure 16b: The area surrounded by a square in Figure 16a –
On the left the original image, in the middle the image after regular filtering and on the right, after geometric filtering according to curvature

Sharpening blurry areas

In order to sharpen a blurry picture, we compute the curvature of every pixel, assuming that pixels with a high value belong to outlines of different objects. The filtering process applied later on changes the value of each pixel according to its curvature so that the obtained picture seems sharper (Namboodiri & Chaudhuri, 2007). Figure 17 below presents on the left a blurred picture, in the middle – sharpening by a classic method of filtering and on the right – sharpening by filtering according to the curvature in every pixel. It is easy to notice the greater sharpness following the geometric sharpening, e.g. in the pupils, hair, and feathers on the hat.

Figuure 17: Sharpening

Segmentation

The state-of-the-art medical imaging machines that exist today around the world encompass algorithms used in geometric methods of image processing and analysis. They allow doctors to analyze these images more accurately and easily than in the past (Angenent et al., 2006). Figure 18 demonstrates a CT image of tissue suspected of being cancerous. The highlighted marking around the tissue is the result of computing the curvature and searching for points with particularly high values of curvature.

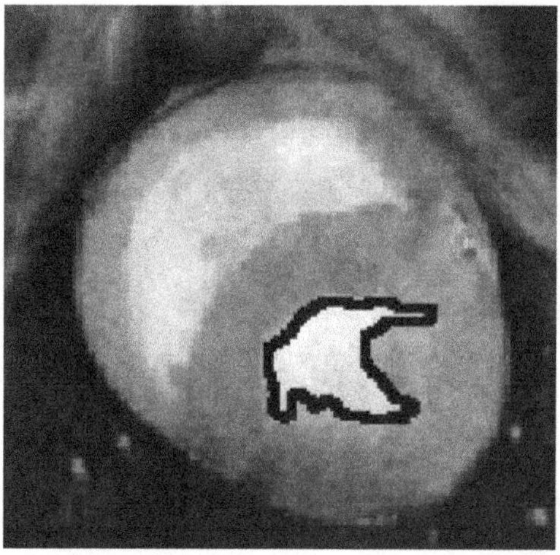

Figure 18: Automatic identification of a tissue suspected of being cancerous in a CT image

Facing the future

The integration of geometric methods and insights in image processing has indeed existed since the beginning of the 1990s. However, this field is still constantly developing. This development transpired simultaneously with the development of theoretical research of geometry that has also accelerated again in recent years. Since the beginning of the 21st century, the application of geometric methods has facilitated the solution of several key problems in mathematics, especially in topology and 3-dimensional geometry. The geometric methods and instruments, developed at the end of the 1990s and beginning of the 2000s for the purpose of solving these problems, are permeating today more rapidly also the various areas of engineering, image processing is one of the first and leading among them. This trend will presumably continue in the near future.

One of the most salient examples in this context is the recently suggested use of a geometric process, namely application of the Ricci curvature function that is more general than the one defined at the beginning of the chapter. This process is called **Ricci Flow,** and at the beginning of 2003, it enabled the solution of two of the difficult mathematical problems that had preoccupied mathematicians throughout the 20th century (Morgan & Tian, 2007). Figure 19 demonstrates a noisy picture in which the noise intensity is particularly high in relation to the picture content. It is difficult to deal with such noise levels even with the geometric methods prevalent until now. The Figure presents the result of filtering the picture by means of Ricci Flow. The result is in fact not as clean as the original picture. However, its quality surpasses the result achieved by previous geometric methods and most certainly by all the classic methods (Appleboim, Saucan, & Zeevi, 2012).

Figure 19: On the left the noisy picture, in the middle the picture after standard geometric filtering and on the right filtering according to Ricci Flow

From Theory to Practice: Proposal for Teaching Non-Euclidean Geometry

I have had a good reason for choosing the issue of image processing as a means of making learners like geometry and acknowledge its relevance and important contribution to fields that are the leaders of industry and technology. The fact that the problems in the field of image processing have not yet been completely solved and that there is room for improvement enhances further its potential to evoke interest and enthusiasm among learners. Nevertheless, teaching this subject obliges us as teachers to cope with the far-from-simple challenge of clarifying the key ideas and notions in image processing and non-Euclidean geometry, preparing learners to explore and think about possible solutions. This part discusses several ways of teaching non-Euclidean geometry through demonstrations and activities, according to the order of the topics reviewed in the theoretical part.

Non-planar surfaces

How can we demonstrate to the class pupils the revolutionary discoveries of Gauss and Riemann? For that purpose, we are going to recruit the important mathematician **William Paul Thurston** (Figure 20). Thurston has been a precursor mathematician and laureate of the Fields Medal for his studies in 3-dimensional geometry and topology. He was endowed with an amazing ability to transmit his most profound mathematical ideas in the simplest and clearest manner. Parallel to pure mathematical research, he extensively engaged in ways by which mathematics should be taught (Conway, Doyle, Gilman, & Thurston, 1991; Thurston, 1994).

Figure 20: Bill Thurston in a geometry workshop for high school students

Thurston used to give the following example in the geometry workshops that he constantly conducted in high schools around the United States until his demise in 2012. The example aimed to demonstrate an essential geometric difference between a planar surface and a non-planar one.

Choose any citrus fruit and draw a circle around any point on its peel. Now peel the citrus fruit so that the peel remains whole (i.e. in one piece). Then spread the fruit peel on any flat surface – floor, table, board. At this stage, think about what has happened to the shape of the peel, paying particular attention to what happened to the circle you have drawn on the peel prior to the peeling. Does the drawn line still complete a circle even after spreading the peel?

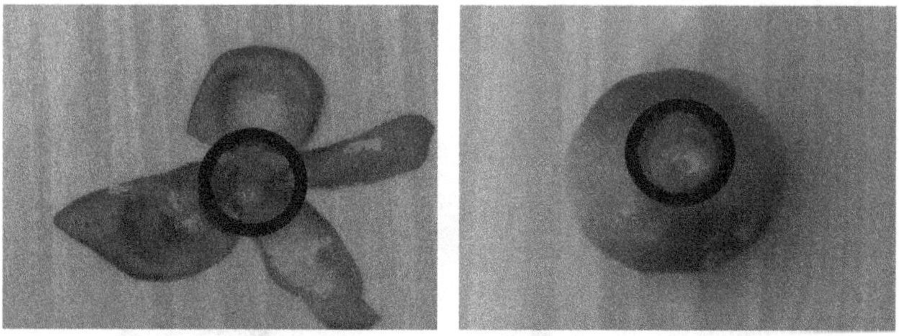

Figure 21: The citrus fruit peel before and after being spread. What happened to the circle?

At this point, we should indicate that fruits like mandarin or tangerine are recommended in particular for the purpose of the example due to their thin and flexible peel. Oranges too can do the job quite well. Conversely, a grapefruit or pomelo are a real challenge reserved only to those who "excel" in peeling the fruit since the peel of these fruits is thicker and less flexible and hence is easily torn. Bon Appetit.

Curvature

Since curvature is a key term in non-Euclidean geometry, it is highly important to consolidate the learners' understanding of it. This is an especially challenging task because it requires a change and expansion of the thinking to which learners have been accustomed. Thus, for example, when we deal with curved surfaces, the known theorem "between two points only one straight line can be drawn" is not relevant. As mentioned, we do not usually relate to the straight line between two given points on a curved surface but to the shortest line between the two points. This can be easily demonstrated with the following example that presents two facts from everyday life:

1. Transatlantic flight paths are planned so that they fly along the shortest course between the departure point and the destination point.

2. The map in Figure 22 shows the shortest flight path between Tel Aviv and New York.

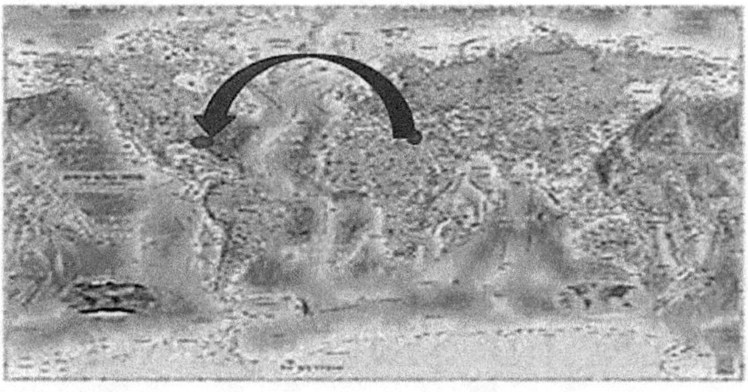

Figure 22: The direct flight path from Tel Aviv to New York:
Is it possible that this path is the shortest between the two cities?

Are these two facts consistent with each other? If we had been asked to draw the flight path from Tel Aviv to New York on the given map, would we have drawn the path indicated in the Figure?

The expected answer is that probably we would not have drawn the indicated path but would have drawn a straight line between the two cities, assuming that this is the shortest flight path between them. The reason that the shortest course between Tel Aviv and New York is the drawn arched course is due to the fact that Planet Earth has a spherical rather than a planar shape. Hence, the drawing of Planet Earth as a planar surface, as it is presented in the Figure, is not compatible with reality. The short lines on a spherical surface are different from the short lines on a plane.

Another difficulty is the need to explain to the students the properties of curvature. Among the five properties of curvature indicated above, there are facts that are harder to demonstrate and understand. However, the following property of curvature - **in a surface whose curvature is positive at any point, the sum of the angles in a triangle is always greater than 180°** - can be easily demonstrated to learners, even to 6th graders. This can be done in the following way:

Our journey starts at the North Pole. We turn south along longitude 0 and continue until reaching the point where the longitude and the equator meet. At this point, we turn westwards (probably to the right) and continue moving along the equator until we complete a distance of exactly one-quarter of the equator. Now we turn northwards (once more to the right) and go north along the longitude which we have reached. We go on until we arrive, hopefully in safety, back to the North Pole.

If we draw the above path, we obtain a spherical triangle, namely a triangle drawn on a sphere (see Figure 23). What is the sum of the angles of this triangle? Is it equal to 180°?

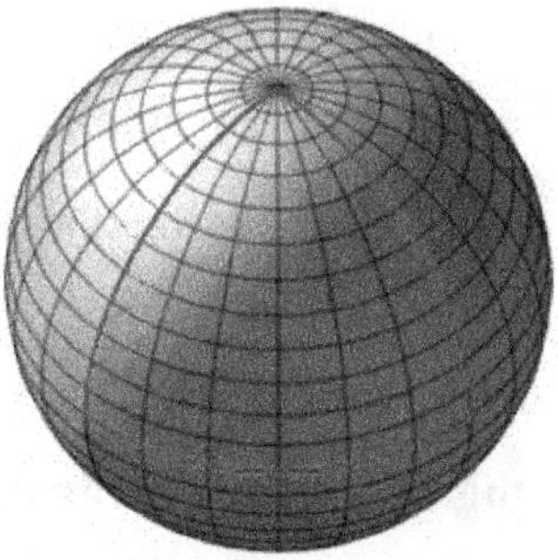

Figure 23: A spherical triangle – What is the sum of the angles?

This new issue should be given to the learners as an inquiry assignment. They can be asked to draw such a path on a globe and measure the angles of the obtained triangle.

Summary from a Personal Perspective

I deem it correct to end the chapter on a personal note. During the academic year 2013–2014, I conducted a workshop in one of the primary schools in the city of Haifa, arranging several meetings with the 6th graders in the school. During the meetings, I presented different mathematical topics that were not part of the school curriculum, the guiding line being applications of mathematics in our daily life (beyond shopping in the neighborhood mini-market). We discussed among others the issue of non-Euclidean geometry. The main ideas were delivered through two examples mentioned in the chapter. The meeting opened by presenting the question about the flight path from Tel Aviv to New York. It was presented as a riddle whose solution would be elucidated during the meeting. At first, the pupils suggested various solutions to the question: “Why is the flight path not the straight line on the map?”, and none of the solutions related to the geometry of Planet Earth. Then I presented the challenge of peeling the citrus fruit, to which the pupils were gladly mobilized. Following the challenge, they slowly started linking the question of the flight path and the shape of Planet Earth. For example, when I asked them why I was using a citrus fruit in order to solve the problem, several pupils answered almost immediately that this was related to the fact that a citrus fruit had a spherical shape like the globe. Hence, I concluded that at least subconsciously they had connected, already at an early stage, the question of the flight path and the geometry of the globe. Later on, the correct answer to the riddle was obtained in a natural and clear manner and without any difficulty.

I feel obliged to indicate the inquisitiveness, enthusiasm, and openness of opinion and thought with which the pupils accepted the contents delivered during the workshop. They were definitely involved in it, even those who did not excel in mathematics studies. I am confident that a large number of pupils are endowed with this natural inquisitiveness. My successful experiment of demonstrating non-Euclidean geometry and the enthusiasm with which it was accepted, illustrate in my opinion that there is room for presenting this topic to varied pupil audiences, at least on an intuitive level.

We are living in an age, whereby we are seeking ways of rendering mathematics in general and geometry, in particular, more appealing to the pupils. Hence, we should try familiarizing them with geometry through its applicable and updated aspects, mobilizing them to solve existing problems without experiencing fear. Learning through numerous examples, illustrations and inquiry tasks will not only be embedded in the pupils' brains. It will also offer them a positive, enriching and enjoyable learning experience, exposing them to the beauty of geometry.

References

1. Angenent, S., Pichon, E., & Tannenbaum, A. (2006). Mathematical methods in medical image processing. *Bulletin of the American Mathematical Society, 43*, 365–396.
2. Appleboim, E., Saucan, E., & Zeevi, Y. Y. (2012). Ricci curvature and flow for image denoising and super-resolution. *IEEE proceedings of European Signal and Image Processing Conference* (pp. 2743–2747). Bucharest: IEEE.
3. Conway, J., Doyle, P., Gilman, J., & Thurston, W. P. (1991). *Geometry and the imagination.* Notices of the American Mathematical Society. Minnesota: Geometry Center, University of Minnesota.
4. Do Carmo, M. (1976). *Differential geometry of curves and surfaces.* New Jersey: Prentice Hall Inc.
5. Gonzalez, R., C., & Woods, R., E. (2008). *Digital image processing* (3rd edition). New Jersey: Prentice Hall.
6. Henderson, D. (2005). *Experiencing geometry: Euclidian and non-Euclidian with history (3rd edition).* New Jersey: Pearson Prentice Hall.
7. Morgan, J., & Tian, G. (2007). *Ricci flow and the Poincaré conjecture*. Cambridge, MA: American Mathematical Society.
8. Mumford, D., & Shah, J. (1985). Boundary detection by minimizing functionals. In S. Ullman & W. Richards (Eds.), *Image understanding* (pp. 19–43). Huston, TX: Ablex Press.
9. Namboodiri, V., P., & Chaudhuri, S., (2007). Image restoration using geometrically stabilized reverse heat equation. *IEEE proceedings of*

International Conference on Image Processing (pp. 413–416). San-Antonio, TX: IEEE.

10. Sochen, N., & Zeevi, Y. Y. (1988). Images as manifolds embedded in a spatial non-Euclidean space. *IEEE proceedings of International Conference on Image Processing* (pp. 166–170). Chicago, IL: IEEE.
11. Spira, A., Sochen, N., & Kimmel, R. (2005). Geometric filters. In C. E. Bayro (Ed.), *Handbook of geometric computing* (pp. 203–230). Heidelberg: Springer-Verlag.
12. Spivak, M. (1999). *A comprehensive introduction to differential geometry* (3rd edition). Texas: Publish or Perish Inc.
13. Thurston, W. P. (1994). On proof and progress in mathematics. *Bull of the American Mathematical Society, 30*, 29–37.
14. Yau, S. T. (2001). Geometry of three manifolds and existence of black holes due to boundary effects. *Advances in Theoretical Mathematical Physics, 5*, 755–767.

https://doi.org/10.1142/9789811218750_0009

Chapter 9

The Concept of Invariance in Geometry: A Long-Term History and Present Day Implications for Learning

Ilya Sinitsky and Moshe Stupel

Introduction: Concepts of Invariance and Change

The concept of invariance expresses people's attempts to determine the situation of something constant on the background of the reality of continuous changes. This notion constitutes a way of finding an "anchor" and comprehending changing events that transpire around us. The concept of invariance has a meaning only when a situation is transformed into another one, whereby some of the characteristics change while others remain unchanged. Consequently, the fundamental concepts of "invariance" and "change" are intertwined and turn up together in various philosophical theories (particularly in the Ontology) and religions.

The search for constant components that enable the description of the physical and mental world, relates to some components of mathematics knowledge either in an explicit or implicit manner. In the classical Greek philosophy, four possible combinations of two states (wet or dry, hot or cold) for two traits of matter constitute four elements of the physical world. Moreover, each one of them is associated with one of the Platonic solids. In Aristotle's Theory of Essence, a fifth, mental element (Quinta Essentia) is added. This attains the match between the fundamentals of philosophy and the "fundamental polyhedra" of solid geometry (Lloyd, 1968).

The concept of number is one of the first and most basic examples of the development of abstract thinking in general and mathematical thinking in particular. It expresses invariance of a quantity that does not depend on the special nature of the counted objects and the way one counts them. Similarly, every basic concept of mathematics is obtained as an abstraction that constitutes the invariance of the specific features of the object while it is being changed in some way. Mathematical concepts relate to the sets (usually infinite) of objects with a characteristic feature, e.g. even numbers, parallelograms, cyclic groups, and others. Yet, changing a characteristic feature entails a change of the object. Determining the possible boundaries of the changes that still preserve the belonging of the object to the specific set, is one of the prevalent approaches in contemporary mathematics. However, even pupils are already familiar with examples that reflect the approach. For example, the fact that each quadratic equation has two real solutions if and only if the value of its discriminant is greater than zero.

Furthermore, mathematical rules, starting with simple rules of the four basic mathematical operations with numbers, express the invariance of an existing state while generating a certain change. Thus, for example, the sum of numbers is maintained when their order is changed; the distance between two numbers on the number axis is maintained when they are both shifted by the same segment in the same direction; the measure of the angle between two rays is maintained in every rotation of the plane and so forth.

Viewing the invariance in sets of objects and in various phenomena leads to a generalization that is consolidated following a single experience and constitutes the core of mathematics knowledge (Mason, 1996). One can identify aspects of invariance in many situations that pertain to school mathematics (Libeskind, Stupel, & Oxman, 2018). In the world of numbers, this is an algebraic view that is not necessarily manifested in the language of letters and symbols (Sinitsky & Ilany, 2016).

The aspects of change and invariance have a long-term history in the development of geometry, which is considered as the first branch of mathematics that was organized as a theory. On the one hand, a view of invariance characterizes deductive geometry from its very beginning. On the other hand, the issue of the possible change and its impact on the features of the geometric objects is a source of geometric theorems and even serves as an instrument for proving them. Hence, the approach of invariance and change has a special place in the world of geometry.

The History of Geometry through the Lens of Invariance and Change

Pre-Euclidean period

Geometric knowledge was conceived in various cultures due to practical needs and encompassed the facts that provided instruments for the measurement of lengths, areas, and volumes. These facts related mainly to particular cases. For example, already in the 3rd millennium B.C., the ancient Chinese and the Babylonians were aware of the fact that the area of the square built on the long side of a triangle with sides of 3, 4 and 5 units of length is equal to the sum of the areas of squares built on the two other legs (see Figure 1). In ancient Egypt, this knowledge as well as additional empirical facts about dimensions, shapes, and solids, were used for practical needs (Struik, 1987).

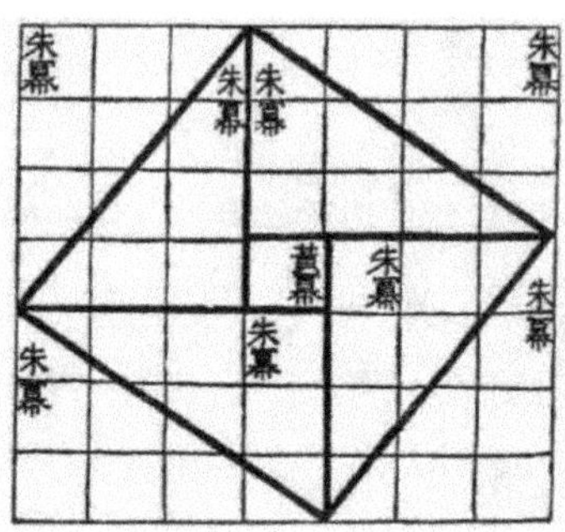

Figure 1: Visual illustration of the Pythagorean Theorem for a triangle whose sides are 3, 4 and 5, in ancient Chinese books (Retrieved from http://commons.wikimedia.org/wiki/File:Chinese_pythagoras.jpg#mediaviewer/File:Chinese_pythagoras.jpg).

Historically, paving the way to geometry as a science is directly associated with the concept of invariance. Apparently, the Babylonians (circa 1500–1000 B.C.) and the Greeks in the times of Pythagoras (in the 6th century B.C.) already knew with a certainty that the equality of the areas mentioned above is valid for **every** right-angled triangle. Moreover, they attempted to construct, in an algebraic way, triples of integers, known today as Pythagorean Triples (Struik, 1987).

The second peak of the pre-Euclidian geometry relates to the findings made by Thales. A famous Thales' theorem states that an inscribed angle subtended by a diameter is a right angle (Figure 2). In other words, changing the location of the vertex of the angle on the circle does not change its measure and it remains a right angle.

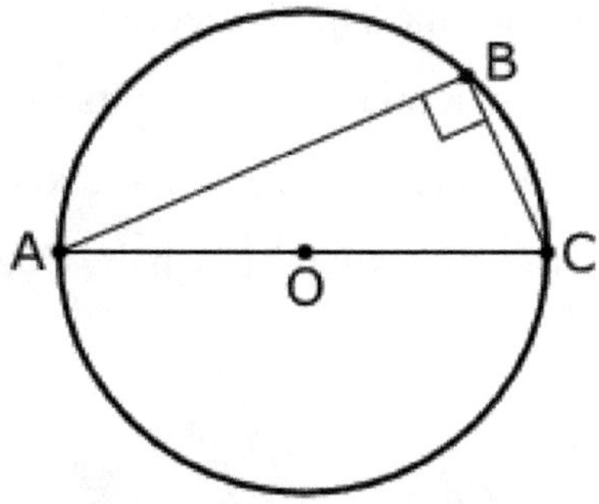

Figure 2: When the point B on the circle is moving, the visual angle of the diameter remains a right angle. (For a dynamic illustration, see http://en.wikipedia.org/wiki/Thales%27_theorem#mediaviewer/File:Animated_Illustration_of_thales_theorem.gif).

Other Thales' theorems (also known as intercept theorems) indicate the invariance of the ratio between the segments intersected by the parallel lines (Figure 3).

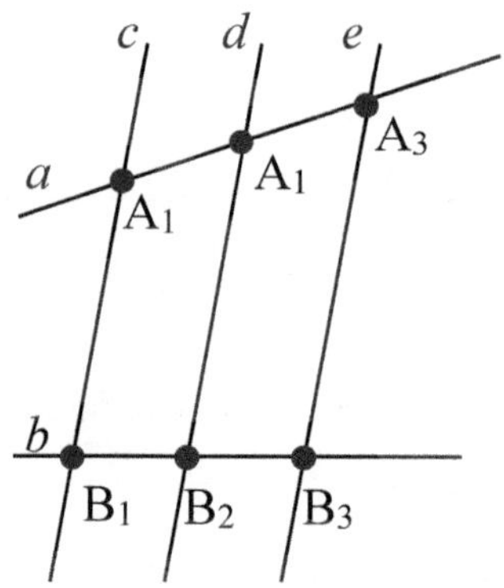

Figure 3: One of the versions of Thales' theorems: when lines *a* and *b* are intersected by parallel lines (*c*, *d*, *e*), the ratios of the segments on the line *a* are equal to the ratios of the corresponding segments on line *b*.

In one of the versions, the Thales theorem states that the construction of segments of equal length on one line results in segments that are equal to each other on the other line. This version is a known method of dividing the segment in any given ratio *a:b* for any pair of natural numbers *a*, *b* (see Figure 4).

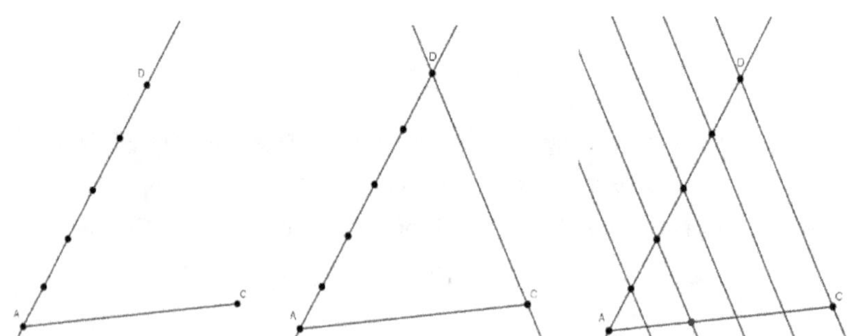

Figure 4: Division of segment AC by a ratio *a:b* according to Thales' theorem: Demonstration of division by a ratio of 2:3.

1. Drawing a straight line through point A and constructing five equal segments on it, thus obtaining segment AD;
2. Joining point D, to point C;
3. Drawing four parallel lines to DC through the point generated in stage 1.

Obviously, the lengths of the equal segments marked on one of the lines can be changed.

Euclid's Elements: The deductive structure of geometry from the aspect of invariance and change

The thirteen volumes of Euclid's *Elements* are considered as the treatise with the strongest impact on the development of mathematics and as the most successful textbook in the history of mankind (Boyer & Merzbach, 1991). The book refers to invariance and change in five major aspects:

- The book is unique first and foremost due to its deductive structure. All the propositions (the majority of which were known before Euclid's time) are grounded in several fundamental assumptions, the postulates or the axioms. For example, "A straight line can be drawn from any given two points regardless of the distance between them" (1st axiom) or "All the right angles are equal" (4th axiom). These postulates are the absolute and invariant truths in every situation of Euclidean geometry. Hence, the theorems derived from the postulates are also absolute and do not depend for example on the size of the geometric objects or on their relative position. It is noteworthy that the word "Euclidean" was redundant for describing this field of knowledge since its conception and until the end of the 19th century. According to mankind's view, the fundamental assumptions of geometry, and therefore its theorems, expressed the entire physical world.

- Euclid stated in a straightforward manner the idea of invariance concerning the sizes and measurements of areas and volumes. In the contemporary notation, the area of every shape is proportional to the second degree of some linear measure, *l*, of the shape, $S = \alpha l^2$, the coefficient α being constant and expressing invariance in the family of similar planar shapes[1]. Similarly, the volume of every solid is proportional to the third degree of its linear measure (namely *l*), $V = \beta l^3$.
- Euclid was well acquainted with equal-area but non-congruent shapes, and he clearly distinguished between equality of areas and congruency. Congruency is formulated in *Elements* in a well-defined language of invariance and change: two shapes are congruent if and only if one can be superimposed on the other by applying the allowed changes, namely isometric transformations of translation, rotation, and reflection.

In his proofs, Euclid distinguished the difference between possible situations that were supposed to emerge while proving the theorems. He usually related to the most difficult case, leaving room for completing the proof in easier cases. For instance, while engaging with the parallelogram area problem, Euclid studied, in particular, the

---

[1]Hence, obviously, the ratio between areas of similar planar shapes is equal to the square of the ratio of the lengths of the corresponding linear elements of these shapes.

most complicated case of partial overlapping of the “shifted” triangles (Figure 5).

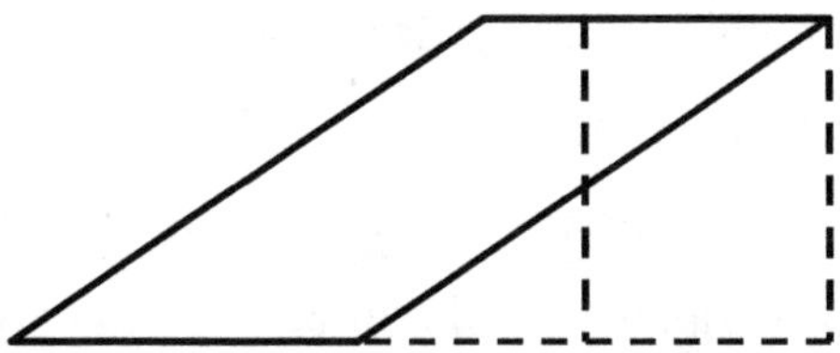

Figure 5: Obtaining a formula for the parallelogram area: The “difficult” case of partial overlapping of the triangles.

- The concept of invariance is systematically applied in *Elements* when proving theorems (not necessarily of geometric theorems[2]). Euclid uses the invariance of the area of shapes obtained by decomposition and re-composition; in the equality of areas of triangles with a common side and equal heights to this side; in the invariance of measure of the inscribed angles with the same circle arc, etc. In order to use the required invariant, Euclid performs auxiliary constructions for finding an invariance that even build and generate the invariance. This aspect is extremely important in the context of the didactic potential of the concept

[2]The book *Elements* engaged also in the theoretical and computational arithmetic that was known to the Greeks in the times of Euclid.

of invariance (several relevant examples are presented further on).

Invariance as an instrument for classifying geometric objects

Another breakthrough in the history of geometry is associated with the systematic use of algebraic language for describing geometric objects. Analytic geometry enabled simplifying the solution of many geometric problems based on a one-to-one correspondence between the plane/space points and ordered pairs/triples of real numbers. The analytical approach establishes numerous invariants for the sets of geometrical objects. Below are two of the most known invariants:

- Every straight line is described by an equation of the first order. In the language of change and invariance, after any isometric or homothetic transformation of the plane, a straight line is transformed into another straight line and its equation in the previous coordinates system is changing. However, the linearity of the equation is maintained.
- All the conic sections are described in analytic geometry by different equations with a common property: they are all of the second order. This generally looks in the following way:

  $$\mathrm{A}x^2 + Bxy + \mathrm{C}y^2 + Dx + Ey + F = 0$$

  (with certain limitations on the coefficient values).

  In other words, the general form of the equation is the invariant for the entire family of conic sections. Moreover, after all the transformations of translation, rotating, reflection or scaling the plane, the equations of each and every curve are obviously changed but their special characteristics are maintained. For example, the sign of the expression $\Delta = B^2 - 4AC$ is an invariance of all these transformations. This invariant also distinguishes the conics: the type of curves can be defined even before deriving its canonical equation (Coolidge, 1968).

These findings have led to a comprehension concerning the importance of the concept of invariance as a characteristic that does not vary under certain changes. Since then, the search for invariants of transformations has embodied an essential approach (and even a leading one) in mathematics, as several mathematicians argue.

Invariance as a characteristic of various geometries

Extensive use of the analytic approach in the 18th and 19th centuries, in parallel to the development of suitable algebraic instruments, facilitated the development of new branches of geometry that explored the internal properties of geometric objects. For example, differential geometry examines the properties of curves and surfaces that are described by analytic functions and portrays the properties of objects in terms of their fundamental characteristics (such as curvature).

Another branch that uses the concept of invariance more prominently is Topology. It engages in the object properties that are maintained under continuous deformations of space[3]. Topology focuses on the object's properties that do not depend on the object's sizes. For example, in elementary topology, there is no difference between distinctive polygons and even between all the simple geometric shapes (bounded by lines that do not intersect themselves). Thus, the topological properties of all these objects are described as invariance under continuous transformations of space.

Towards the middle of the 19th century, Euclidean geometry was no longer the single geometry. Already in the 17th century, a projective geometry started developing, focusing on the properties of curves that are maintained under the projection of the curves from one plane to another. The development of projective geometry was inspired by the wish to comprehend the laws of perspective for the purpose of painting and architecture. However, this progress gave rise to discussions about the essence of invariance of lengths and angles in Euclidean geometry. At the same time, after long-range attempts to prove Euclid's 5th postulate, the parallel postulate, mathematicians succeeded to construct other geometric deductive theories, in which Euclid's 5th postulate was substituted by another postulate.

[3]The first problem associated with topology is Euler's analysis of the Seven Bridges of Königsberg. The problem requires finding a specific way of visiting all the city parts that are connected by the bridges (for additional details see http://en.wikipedia.org/wiki/Seven_Bridges_of_K%C3%B6nigsberg). Discovering a simple invariance of a parity offered Euler a proof for the non-existence of the required path.

This entailed the need for understanding the relation between these new deductive theories and the Euclidean geometry within the collection of various geometries. In 1872, Felix Klein (2004) suggested an elegant novel idea, classifying geometries by their basic symmetry groups, namely the group of all transformations under which the object is invariant. The Euclidean geometry, for example, allows isometric transformations and engages in the object properties maintained under translation, rotation, and reflection. Angles, areas, and distance between the points, are maintained under these changes. Conversely, transformations of projective geometry do not maintain distances and angles but do maintain other special relations. The meaning of the basic concepts of geometry can change as a result of the transition to another geometry. Thus, objects and facts that are very familiar to us from Euclidean geometry, are interpreted differently in projective geometry. For instance, because of non-preservation of angles, the concept of parallel lines does not exist (and from an "applied" point of view, two parallel lines meet at a point at the projective plane). Because the set of allowed transformations of Euclidian geometry is a proper subset of transformations of projective geometry, all the projective geometry claims are valid in Euclidean geometry, but not vice versa.

All the above illustrates that invariance has become a basic concept in the comprehension of the relations between different geometric theories. The interplay between changes (transformations) and invariance is one of the major issues in contemporary geometries.

Theorems in Geometry as "a Game" of Seeking the Invariance and Effect of the Change

View of invariance is one of the important competencies that should be developed in learners. Every geometric theorem can be formulated in terms of invariance and change (Sinitsky & Ilany, 2016, p. 286). Nevertheless, there are theorems in which invariance explicitly plays a prominent role both in their formulation and in their proof.

The invariance of area and its application: Euclid's proof of the Pythagorean Theorem

The Pythagorean Theorem is one of the most famous and basic theorems in the history of geometry (Struik, 1987). The theorem has hundreds of known proofs of various types (Bogomolny, 2015). The following is Euclid's proof based on the invariance of the area of the triangles that have a common side and the third vertex of the triangle moving on the straight line that is parallel to this side (Figure 6).

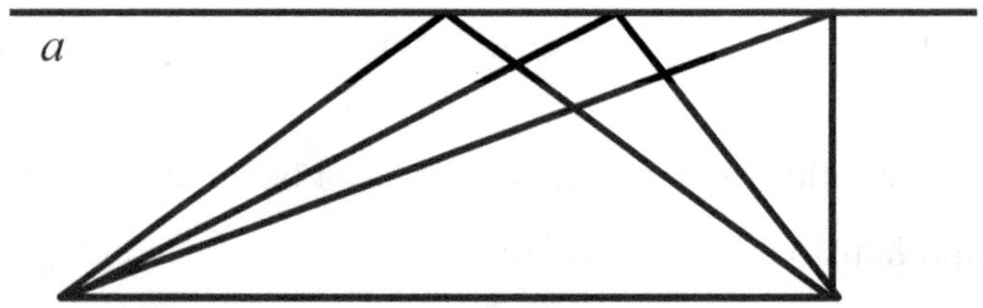

Figure 6: Invariance of the areas of triangles with a common side and a third vertex on the line *a* that is parallel to this side.

All the four triangles with the common side have the same height, hence their area is maintained despite the change of position of the third vertex on the straight line *a*.

Theorem 1: Pythagorean Theorem: The sum of the areas of the two squares built on the legs of a right-angled triangle is equal to the area of the square built on the hypotenuse of that triangle.

Euclid's proof is demonstrated in Figure 7. Segment CG divides the square built on the hypotenuse into two rectangles. The area of triangle ACI equals the area of triangle AFI since points C and F are on a straight line parallel to the common side AI of these triangles. Hence, the area of the triangle ACI is half the area of rectangle AFGI. Similarly, the area of triangle AEB equals the area of triangle ACE: their vertices C and B are on a line parallel to their common side AE. Consequently, the area of triangle AEB is half the area of square AEDC. However, triangles AEB and ACI are congruent (according to the congruency of triangles SAS theorem). Thus, the area of square AEDC equals the area of rectangle AFGI.

Similarly, the area of the rectangle built on the other leg equals the area of rectangle BFGH. The sum of the areas of the small squares built on the legs of triangle ACB is equal to the sum of the rectangle areas, namely to the area of square ABHI.

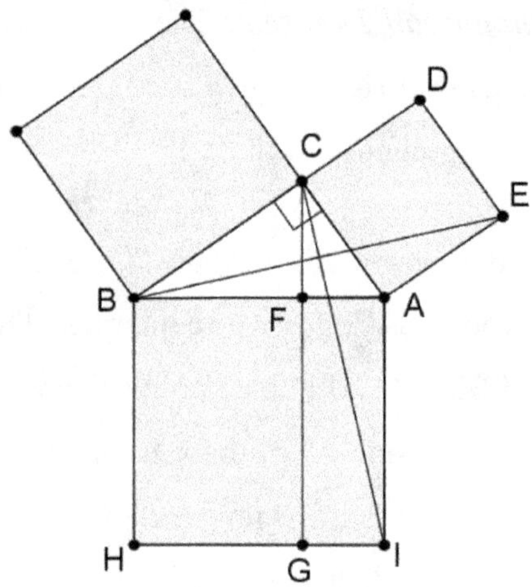

Figure 7: Euclid's proof of the Pythagorean Theorem.
The proof is based on the invariance of the triangle area with the given side and the constant value of the height to this side.

Invariance of orthogonality

As a generalization of Thales' theorem of the invariance of the visual angle of the diameter from a point on the circle, a more general invariance was proven in Euclid's book: equality of inscribed angles on the same circle arc. When the arc is half the circle and the matching central angle is a straight angle, then the inscribed angle is a right one. The following theorem also relates to the invariance of orthogonality.

Theorem 2: Perpendicularity of the tangent line to the radius at the point of tangency: A tangent from point B to a circle with center O is perpendicular to radius OA at the point of tangency (Figure 8).

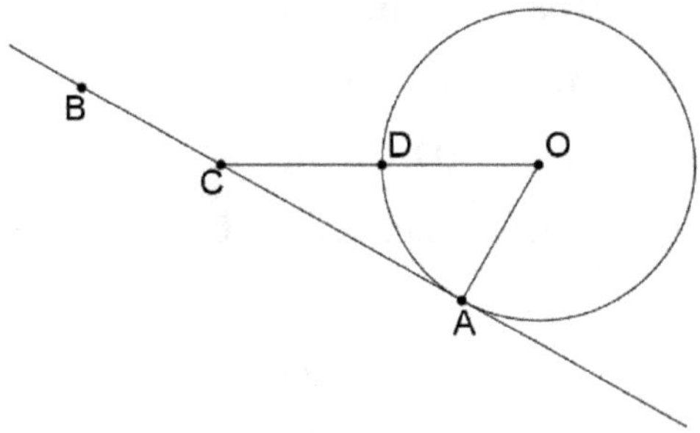

Figure 8: Perpendicularity of a tangent line to the radius at the point of tangency.

The proof of this claim is based on the extremum property of the perpendicular from a point to a straight line: It is shorter than any other segment that connects the point and the line. The distance of any point C on the tangent (that is different from the point of tangency A) from O (CO) is greater than the distance between A and O (AO) ($CO > DO = AO = R$). Hence, segment OA is shorter than any other segment that connects point O with straight line AB and, as a result, is perpendicular to it.

The following formulation of the theorem emphasizes the invariance: changing the point of tangency maintains orthogonality between the tangent and the relevant radius.

Invariance of parallelism

In Euclidean geometry, there are many theorems that relate to the spatial relation between two lines, i.e. the invariance of parallelism theorems. Below we present one of them.

Theorem 3: Parallelism of a line formed by the intersection of "secondary" angle bisectors to the triangle side. If AM is a median to side BC of the triangle, and lines MD and ME are bisectors of angles AMB and AMC, respectively, line DE is parallel to BC, BC || DE (see Figure 9).

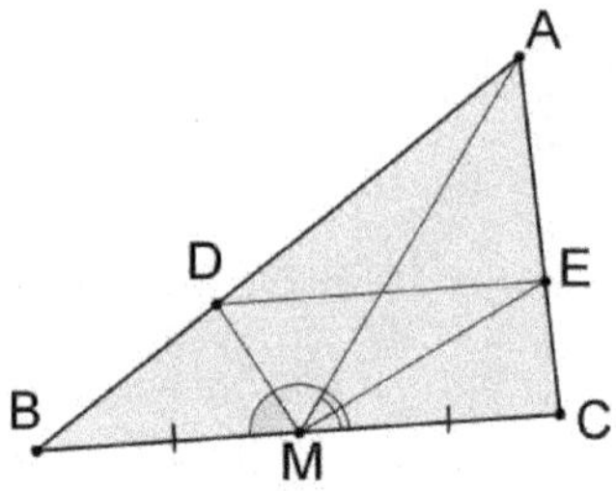

Figure 9: Line DE, connecting the points of intersection D, E of the angle bisectors from point M (middle point of BC), is parallel to BC (BC || DE).

The proof: according to Thales' theorem demonstrated in Figure 4, the angle bisector divides a side into segments in the ratio that is equal to the ratio of the length of the sides adjacent to these segments. That is, the proportions $\frac{AD}{DB} = \frac{AM}{BM}$ and $\frac{AE}{EC} = \frac{AM}{MC}$ refer to triangles AMB and AMC, respectively. Since M is a midpoint of BC, segments BM and MC are equal, implying the equality $\frac{AD}{DB} = \frac{AE}{EC}$. According to the converse of Thales' theorem, BC||DE[4].

---

[4]Dynamic demonstration of the invariance of parallelism is presented in the Geogebra software application in the following link: http://geogebratube.org/student/m556841.

Invariance of a shape

This section ends with a demonstration of invariance of a shape in a case where the conditions of forming the shape are changed considerably. In addition to the invariance of the shape, another invariance is revealed, as will be elaborated below.

Theorem 4: Type of shape obtained by intersecting the angle bisectors of a parallelogram. Parallelogram ACDE is given, with a length of adjacent sides $AE = CD = b$ and $AC = DE = a$ $(b > a)$. The points of intersection of the angle bisectors of the parallelogram are the vertices of quadrilateral MLBN (Figure 10). Does this quadrilateral have some special properties?

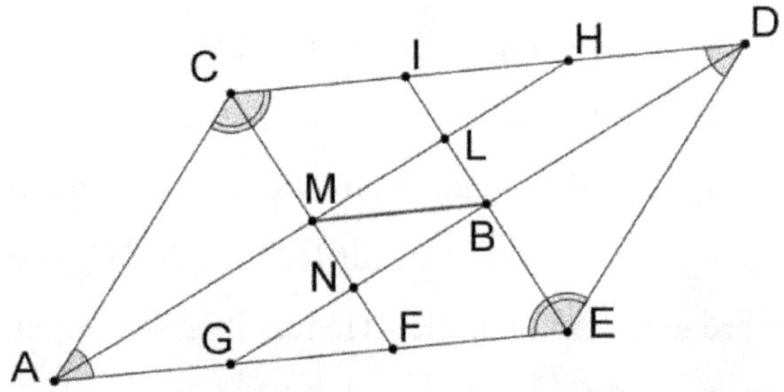

Figure 10: A quadrilateral formed by intersection points of the angle bisectors of the parallelogram.

The sum of two parallelogram angles $\angle$EAC and $\angle$ACD, that are adjacent to the same side, equals 180° (since AE || CD). Thus, the sum of the halves of these angles, namely $\angle$MAC and $\angle$ACM, equals 90° and triangle AMC is right-angled with right angle M. Moreover, angles $\angle$LMN and $\angle$AMC are vertically opposite angles and hence it is also a right angle. Similarly, all the angles of quadrilateral MLBN are right and therefore MLBN is a rectangle. This remains the same for any of the given parallelograms despite its sides and angle measures (except for one special case explored later in Assignment 4).

In this construction, many additional parallelograms can be identified in addition to the given parallelogram and to rectangle MLBN. First, the angle bisectors of the opposite angles of the parallelogram are parallel to each other (for instance, due to the equality of the corresponding angles). Consequently, quadrilaterals AHDG and CFEI are parallelograms. Furthermore, from the proof, it follows that the angle bisector CM is also the height in triangle ACH and, therefore, ΔACH is isosceles and CM is also a median. Thus, the length of segment AM is half of AH. Similarly, the length of segment GB is half of GD, but AH=GD as opposite sides of the parallelogram. Therefore, in quadrilateral AMBG, the opposite sides in every pair are equal and parallel. Hence, it is also a parallelogram and segments MB and AG are parallel, MB||AG. Similarly, quadrilateral MHDB is a parallelogram as well. We attain another interesting invariance: a diagonal of the rectangle MLBN is always parallel to the sides of the given parallelogram[5].

The Concept of Invariance and Change as a Didactic Instrument

Mathematics educators may exploit the explicit wording of various geometric theorems in terms of invariance as well as the remarkable appearance of invariants in numerous situations to encourage learners to search for invariance under various conditions. From this aspect, dynamic instruments such as Geogebra facilitate diverse discoveries and hypotheses concerning invariance of geometric objects, their properties, and measures. In this respect, learners perceive the concept of invariance as an instrument designed for describing mathematical objects and phenomena.

Extensive use of dynamic instruments evokes a discussion about the role of invariance in proofs that are presented as generic examples that emerge following the use of technological tools (Baccaglini-Frank, Mariotti, & Antonini, 2009). In addition to being an effective means of identifying objects and describing their properties, the concept of invariance and change also plays a didactic role. In recent years,

---

[5]Demonstration of this invariance can be seen in the following link: http://geogebratube.org/student/m556837.

mathematics educators have witnessed the development of theories engaging in the ways of exploring the concept while learning geometry (Leung, 2014) and mathematics in general (Sinitsky & Ilany, 2016). In many cases, applying the approach of invariance and change paves the way to the solution of problems and proof of theorems. Examples of an "interplay" between invariance and change in geometric problems are presented later in this chapter.

Understanding the effect of possible change as a means of solving problems

Quite frequently, learners know a specific solution of a problem or a partial solution thereof and they have to find the complete or general solution. Understanding the effect of the possible change can provide a short and effective solution in these cases, as demonstrated by the following assignments.

Assignment 1: Parallelograms with the same set of segments. The length of the adjacent sides of the parallelogram in Figure 11(a) are a and b units of length. Are there any other parallelograms with the same side lengths? Among all these parallelograms, which one has a maximal perimeter and which one has a maximal area?

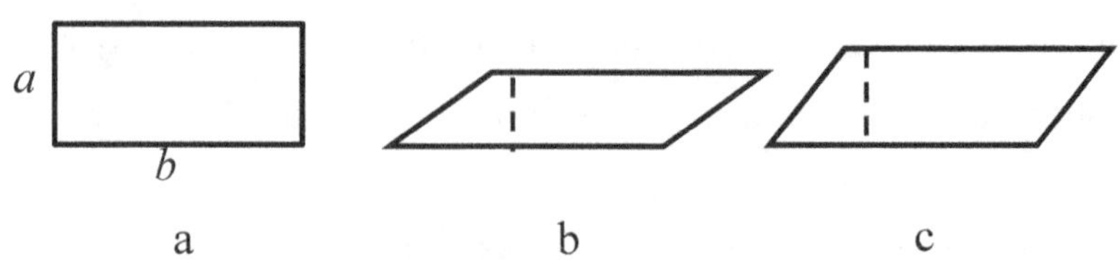

Figure 11: Different parallelograms formed by one set of segments.

Obviously, this assignment is well known. However, we present here the analysis of the problem in terms of invariance and change. Let's start with the following question: "According to the conditions of the problem, what is variant and what is invariant?" Given that the length of the parallelogram sides should not be changed, the only possible change is the measure of the angles between the sides. We can generate an infinite number of parallelograms, all have the same side lengths, and hence, the perimeter does not change. On the contrary, the area does change, leading to the question: "What is the effect of changing the parallelogram angles on its area?" Since the parallelogram area is the product of the side length and the length of the height to this side, let's focus on this product. The first factor is constant whereas the second factor changes with the change of the parallelogram angles. When the acute angle between the adjacent sides is increased, the length of the height is gradually growing, until the angle is no longer acute but right. An additional increase of the angle entails a shortening of the height (see Figure 11). Thus, among parallelograms with the same sides, a parallelogram with a maximal area does exist, and it is a right-angled parallelogram. In other words, with the same two pairs of equal segments, one can compose many parallelograms with the same perimeter and, among these parallelograms, the rectangle has the maximal area.

Assignment 2: A tessellation problem (Sinitsky & Ilany, 2016, p. 94). A tessellation is the tiling of a surface using geometric shapes, with no overlaps and no gaps. Filling the plane with congruent squares (Figure 12a) is the most known tiling type. Two additional tiling types are also well-known: tiling with congruent regular triangles (Figure 12b) and tiling with congruent regular hexagons (Figure 12c). Prove that no additional tiling can be formed by means of congruent regular polygons.

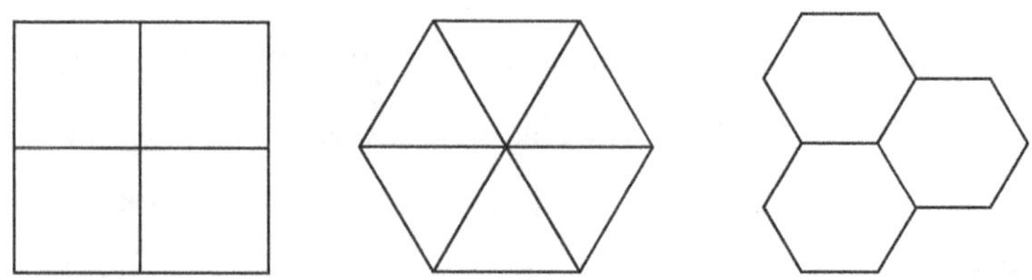

a b c

Figure 12: Various options of tiling with congruent regular polygons.

This statement is usually proved by a relatively simple algebraic technique: in order to ascertain that the plane is filled, the angles of the polygons that meet at every junction add up to 360^0. It leads to one linear equation with two unknowns – the number of sides of the polygon and the number of polygons that meet at every vertex. Because both values are natural numbers, the number of solutions is finite. However, let's look at the most known tiling type (Figure 12a) and try changing it. First, we decrease the number of sides of the tiling polygon and thus, tiling with the regular triangles is obtained (see Figure 12b). A further decrease in the number of sides is impossible since three is the minimal number of sides of a polygon. Then, let's increase the number of sides of the initial polygon suggested for tiling. This increase is accompanied by the increase of its angles, resulting in a smaller number of angles that fill each junction. Consequently, for the tiling with a regular polygon of more than four sides, the number of such polygons should be smaller than for the case of tiling by squares, namely less than four. This can be done by tiling with three polygons, that are necessarily hexagons (as demonstrated in Figure 12c).

In the case of three other congruent regular polygons that are not hexagons, three angles that meet at the vertex do not produce add up to 360^0 at the junction point. A further decrease of the number of polygons meeting at the tiling vertex is impossible: two inner angles of a convex polygon do not add up to a full angle, so there is no tiling with two congruent polygons at the vertex. Therefore, when tiling with regular polygons, the restrictions of the problem do not allow tiling types other than those illustrated in Figure 12.

Discovering the invariance as a way of solving a problem

As indicated above, every theorem can be formulated in terms of invariance and change. In many cases, the discovery of a hidden invariance leads to the ultimate solution of the problem or offers a meaningful stage in its solution. In what follows, we demonstrate this approach through various types of geometric assignments: two extreme value problems concerning the comparison of perimeters and comparison of areas – and a geometric construction problem.

Assignment 3: Shapes on the squares board. The polygon in Figure 13 consists of five square-units. Suggest ways for calculating its perimeter. Is there a different polygon on this squares board, that consists of the same number of square-units but has a different area? Is there a different polygon on this squares board, consisting of the same number of square-units but has a different perimeter? What is the maximal perimeter and what is the minimal perimeter for all these polygons?

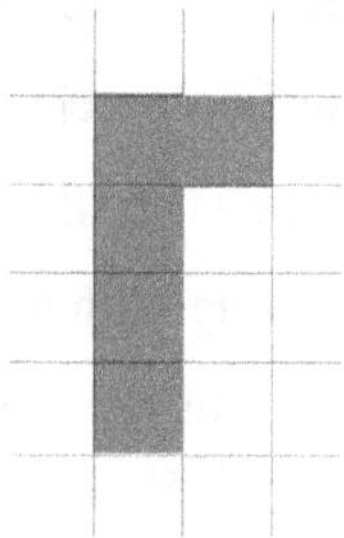

Figure 13: A polygon consisting of five squares drawn on a squares board.

According to the principle of area invariance, all the shapes assembled of the same number of equal squares, have the same area (since these squares are attached to each other with no overlapping).

It is rather easy to find the perimeter of the polygon by counting the perimeter segments. However, it is advised to identify on a more efficient way for determining the perimeter, particularly when searching for the maximal and minimal perimeter among shapes comprised of five square-units.

The perimeter of the polygon in the Figure 13 is 12 units of length. Let's try forming another polygon from the same five squares. An immediate accessible example is a rectangle ranging over five adjacent squares (Figure 14a). Its perimeter is 12 units of length, like the polygon in Figure 13, and this is also the perimeter of several other polygons consisting of five square-units (Figures 14b–14d).

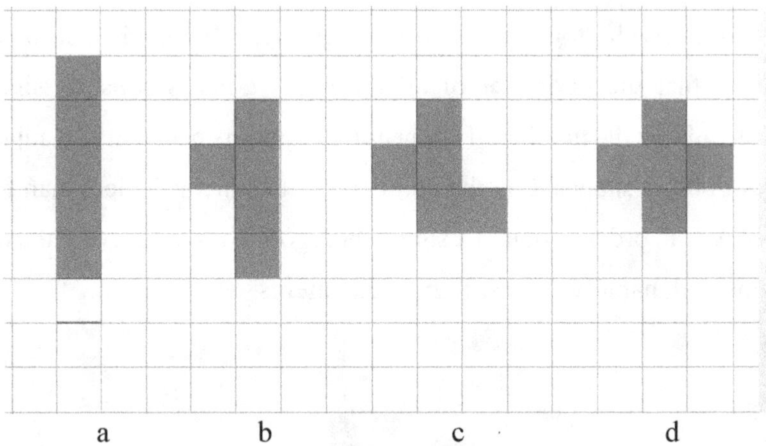

Figure 14: Polygons consisting of five squares, the perimeter of which is identical to the polygon in Figure 13.

At this point, it seems logical to ask the following question: "Can the perimeter of the polygon consisting of the five square-units on a square-lattice board change at all or all these polygons necessarily have the same perimeter of 12 units of length?" The assumption that the answer is positive is based, for example, on the fact that all polygons consisting of three square-units on the squares board (namely, two such polygons) do have the same perimeter.

Another question is raised: "Why do all the polygons consisting of five squares in Figure 14 have the same perimeter?" It is easy to realize that the rectangle in Figure 14a is obtained from the initial polygon by a rotation of one square-unit. Instead of being attached to the adjacent square "on the side", it is attached "from the top". Similarly, the shape in Figure 14b is also different from the given polygon only in the place of joining this square-unit. In fact, all the mentioned shapes are formed by five square-units with four connecting segments. It seems reasonable, therefore, that the perimeter of all these polygons is constant since it depends only on the number of connecting segments between two adjacent squares in these shapes. Can the number of attachments be less than four? No, because in order to obtain a simple polygon, all five square-units must be connected, namely at least four "attachments" are required.

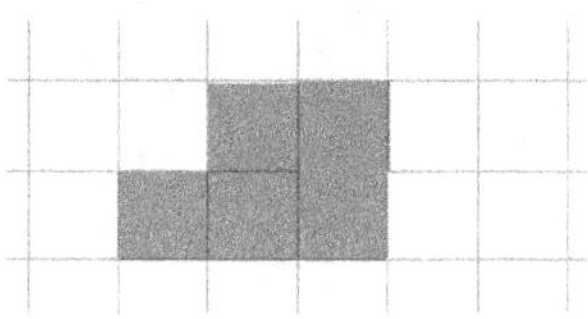

Figure 15: A "special" polygon consisting of five squares drawn on the squares board: There are five "attachments" between the squares.

Can the number of attachments be greater than 5? An attempt to compose a shape of five square-units with more than four attachments results in the polygon demonstrated in Figure 15. Its perimeter is shorter than the perimeter of the previous polygons and is equal to 10 units of length. An extra "attachment" reduces the perimeter by two units of length. The reason for that is the invariance demonstrated in Figure 16.

a b

Figure 16: Perimeter of a shape consisting of two squares (16a) is two units of length less than the sum of the perimeters of two separate squares (16b).

In general, the perimeter of the shape on the squares board is equal to the sum of the perimeters of all the squares minus twice the number of attachments (Sinitsky & Ilany, 2016, p. 357). The perimeter of the m squares polygon with k attachments equals to $P = 4m - 2k,$ where the minuend is the perimeter of every m -squares "separated" from each other and the subtrahend expresses the reduction of the perimeter as a result of k attachments. There are exactly two options for polygons with five squares: all the polygons in Figure 14 comprise four attachments and hence they have an equal perimeter of 12 units of length. Conversely, the polygon in Figure 15, with a maximal number of five attachments, results in a minimal perimeter of 10 units of length.

Assignment 4: Searching for a rectangle of a maximal area. As shown in the discussion of Theorem 4 above (following the proof that quadrilateral MLBN is a rectangle), the quadrilateral whose vertices are intersection points of the angle bisectors of the parallelogram, is a rectangle and its diagonals are parallel to the sides of the parallelogram. Among the parallelograms in which the lengths of the adjacent sides are a and b units of length ($b > a$), in which of them the area of the obtained rectangle is maximal and what is its area? (See Figure 10.)

Similarly to Assignment 1, the lengths of the sides of these parallelograms are constant, and only the angles between the sides can change. Accordingly, the obtained rectangle is obviously changed. And what about its diagonals? The angle between them changes because it is equal to the angle between the parallelogram sides. But what about the length of the diagonals? Let's go back to Figure 10.

In the discussion concerning Theorem 4, we pointed out that diagonal MB of the rectangle is parallel to the sides of the parallelogram. Hence, each of the two quadrilaterals AMBG and MHDB is a parallelogram and the length MB is equal to length AG. But what is the length of AG? Segment DG is an angle bisector, angles ∠GDE and ∠GDC are equal. Moreover, angles ∠DGE and ∠GDC are a pair of alternate angles by intersecting the parallel sides of the parallelogram. Consequently, angles ∠DGE and ∠GDE are equal and ΔDEG is an isosceles triangle. Thus, the length of GE is equal to the length of ED, $|GE| = |ED| = a$ and length AG is the difference between the length of the sides, $|AF| = b - a$. In the application mentioned above (http://geogebratube.org/student/m556837), we can see what happens when the lengths of the bases get closer to each other: the rectangle becomes smaller until the angle bisectors coincide. In the case of $a = b$, namely when the given parallelogram becomes a rhombus, the rectangle entirely disappears.

Let's formulate now this result as an invariance: all the rectangles obtained by changing the angle between the sides of the parallelogram are rectangles with an invariant length of their diagonals. Among all these rectangles, the maximal area attained when the diagonals are orthogonal (this can be seen for example in a formula for the calculation of the rectangle area through using the length of its diagonals and the angle between them). However, the fact that the diagonals of a rectangle are orthogonal implies that it is a square. The area of this square is a half product of its diagonals, $(b-a)$ each one, so the maximal possible area of the rectangle is $S_{max} = \frac{(b-a)^2}{2}$ and it is attained when the rectangle is a square. Moreover, this square has also a maximal perimeter among all the rectangles obtained in the construction described above. It is noteworthy that these extremum features are obtained for the "inner" rectangle when the angle between parallelogram sides is a right angle, in other words, the parallelogram itself becomes a rectangle.

Assignment 5: The problem of the bridge-building. Points A, B are on two sides of a road. The points should be connected in the shortest way.

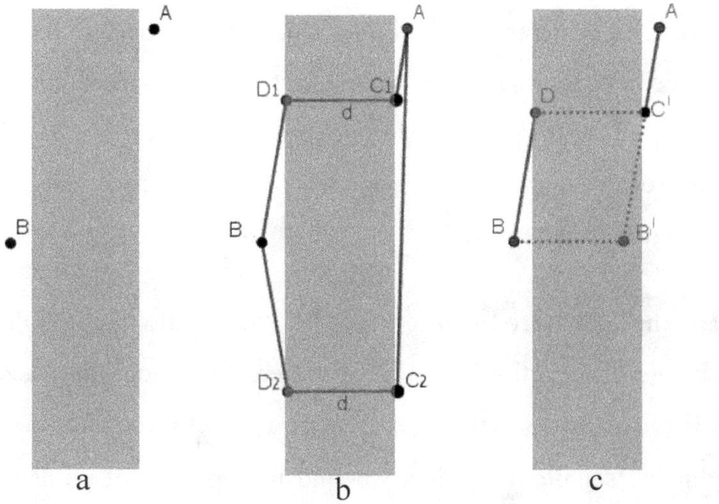

Figure 17: Problem of building the bridge (a) and its solution (b, c).

This way should include the bridge that is perpendicular to the road (Figure 17a).

The problem deals with the choice of two points, C and D, that are positioned on the same perpendicular to the road shoulders and the segment connecting them is the planned bridge. Points A and B are supposed to be connected to points C and D (respectively) in of straight-line segments (since this is the shortest line between any two points). Let's choose randomly two such paths and try comparing their length (Figure 17b). We cannot know which of them is shorter but we can see that both consist of three types of segments: the segment connecting point A with point C_i (either C_1 or C_2) on the road shoulder; the segment C_iD_i (namely C_1D_1 or C_2D_2) of the bridge that connects the road shoulders; and the segment connecting between point D_i (D_1 or D_2) and point B. In spite of the possible difference between the length of segments AC_i and the possible difference between the length of segments BD_i, the length of segment C_iD_i is invariant – it is the length of the bridge, *d.* Hence, the difference between the lengths of the different paths does not depend on the length of the bridge. The shortest path along a certain bridge will be maintained as the shortest path among the built paths in the case of any other length of the bridge.

Having discovered that, completing the solution is straightforward. We change the length of the bridge to zero (no bridge) and accordingly point B moves to its new place, B′. We connect these two points with a straight-line segment and thus we find the position of point C′ on one shoulder of the road. This is the location of the bridge that constitutes the shortest path between A and B (Figure 17c).

After finding the invariance, the problem solution includes a certain change that maintains the difference between the lengths of the paths. The idea of producing the change that maintains a certain property is demonstrated in detail in the following section.

Constructing the change as a means of solution

Assignment 6: Constructing an angle bisector. An angle bisector should be constructed without touching the vertex of the angle (Figure 18).

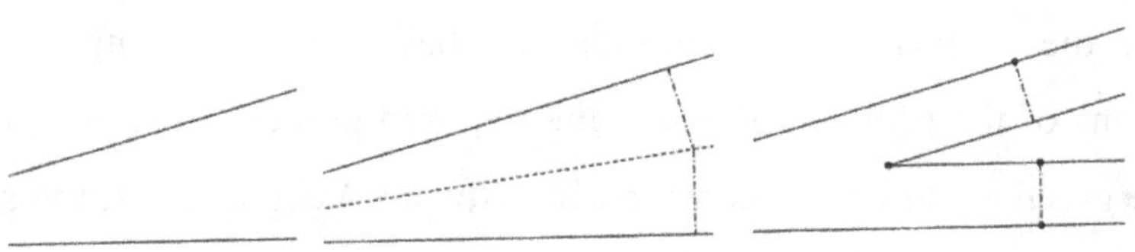

a b c

Figure 18: The problem of constructing an angle bisector without touching the angle vertex.

Let's start the problem solution with a typical assumption for such problems. We assume that the required line that is the angle bisector has already been constructed (see Figure 18b). What characterizes the angle bisector? By definition, it divides a given angle into two identical angles. Moreover, according to the main property of the angle bisector, every point A on it is equidistant from the two rays of the angle (see Figure 18b). Had the angle rays met, it would be possible to easily construct the required line. How can we change the position of the lines forming the angle so that the line that we are seeking (the dotted line in Figure 18b) maintains the property of equidistance from the two lines? The change is quite obvious: each of the lines should be shifted by the same distance in parallel to itself. Moving both lines by the same distance reduces their distance to the required line by the same value and the equidistance up to the new lines guarantee equidistance up to the given lines. The required construction is illustrated in Figure 18c: following the translation to the same distance, the vertex of the angle is already available and the angle bisector of the obtained angle is exactly the required line, i.e. the angle bisector of the original angle.

For the purpose of the solution, we have made a change in the conditions of the problem, keeping the required property: the new angle and the given angle have a common bisector. In Assignment 7, the same idea is applied in order to prove a property that in itself expresses a nice and rather surprising invariance.

There are many theorems that deal with the invariance of sums of certain segments in a polygon. For example, the invariance of the sum of distances from a certain point in a regular polygon to its sides (see Figure 19 that demonstrates the cases of a regular triangle and a square). The following assignment also engages in the invariance of the sum of segments.

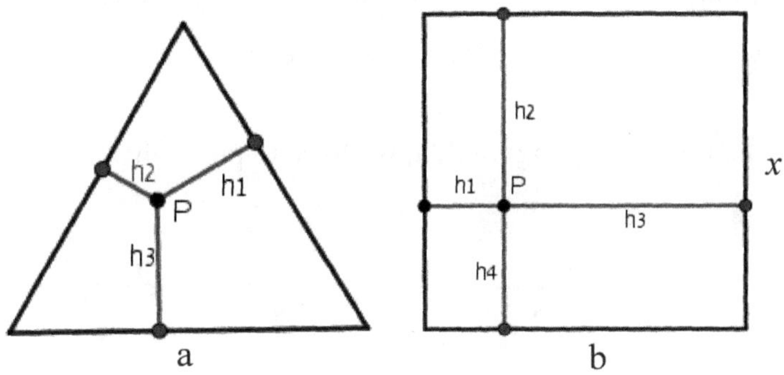

Figure 19: The sum of distances from arbitrary point P to the sides of the two regular polygons is constant for each of them.

In Figure 19a, the case of the equilateral triangle, the sum of these distances is equal to the height of the triangle *h*

$$h = h_1 + h_2 + h_3$$

In Figure 19b, the case of a square, the sum of these distances is equal to twice the length of its sides:

$$2x = h_1 + h_2 + h_3 + h_4$$

Assignment 7: Invariance of the sum of the segment length

Square ABCD with side length a is given.

A straight line is drawn through vertex A, intersecting side DC at point E.

The angle bisector of ∠EAB is drawn through vertex A, intersecting sides BC at point F (see Figure 20a). Prove that BF + DE = AE.

In the notation of Figure 20a, we have to prove that $z = x + y$.

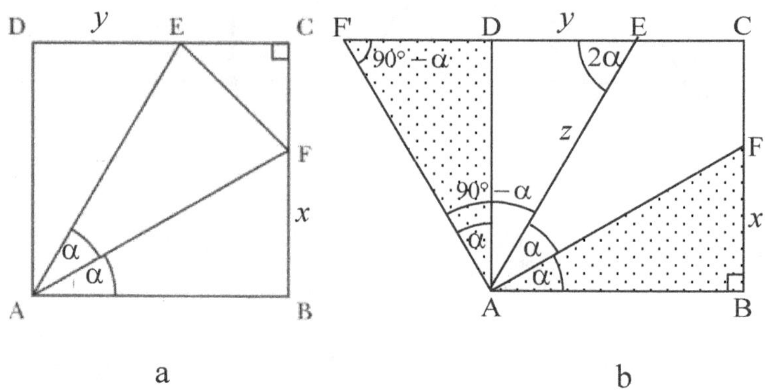

Figure 20. The sum of lengths of segments on squares' side is equal to the length of the section segment, $BF + DE = AE$.

(a) Dissecting a side CD of square ABCD by the arbitrary straight line from the vertex A with further construction of angle bisector.

(b) Proving the identity $BF + DE = AE$ by rotation of triangle ΔABF.

The problem seems complicated since segments FB and DE are neither part of the same segments nor belong to the same triangle. However, it is possible to make some change that keeps one of the segments, say FB, but "combines" it with segment DE.

We rotate ΔABF by 90° in such a manner that side AB coincides with side AD (see Figure 20b). Thus, $F'D$ is equal to FB, $F'D = FB = x$.

Triangle $\Delta F'EA$ has two equal angles: $\angle EF'A = \angle E\ A\ F' = 90° - \alpha$, and it means that $F'E = AE$, or in our notation, $z = x + y$.

Changing the position of segment FB yet maintaining its length, facilitates proving the required property. Note, that the equality between the length of segment AE and the sum of lengths of segments FB and DE is invariant of the direction of the line AE. Moreover, this property is maintained even if the straight line AE does not intersect the side of the square but rather its extension (see https://www.geogebra.org/m/VnpDP).

A similar idea can be applied to proving Ptolemy's Theorem (Sinitsky & Ilany, 2016, pp. 372–374).

Summary

The concept of invariance has a long-term history in the development of geometry. The entire deductive structure of Euclidean geometry is a collection of theorems that engage in invariance under certain conditions of change. Searching for invariance is a key topic also in contemporary geometry. The interplay of invariance and change describes relations within and between geometric objects and their sizes thereof. This chapter presents several theorems concerning the invariance of properties or geometric sizes. Mathematics educators should familiarize learners with these theorems and facts, and highlight the role of invariance embedded in them, including a reference to the possible extent of changes that keeps the invariance.

The wide use of the concepts of invariance and change might be an effective way of discovering the properties of geometric objects and facts that are unfamiliar to the learners. Dynamic geometry environment seems a relevant platform for students' experimentation and thus they become active learners. The examples presented in this chapter relate to issues in geometry studied at different levels. They illustrate that comprehension of the aspects of invariance and change provides an instrument for solving various geometric problems. Understanding the effect of change, discovering the invariance and generating the appropriate change, frequently pave the way to the solution. Nevertheless, discovering the invariance and examining it only with the set of examples cannot replace the formal proof of the theorems that refer to a certain invariance.

The most important didactic aspect of the concepts of invariance and change is associated with strategies for seeking new theorems by the learners. Implementation of this approach with technological instruments, such as dynamic geometry, might considerably influence all the stages of learning, its style and maybe even a paradigm of learning process. This issue is worthy of another study. In one way or another, we trust that the concept of invariance can facilitate more extensively the development and structuring of geometric knowledge of learners at all stages of their studies.

References

1. Baccaglini-Frank, A., Mariotti, M. A., & Antonini, S. (2009). Different perceptions of invariants and generality of proof in dynamic geometry. In M. Tzekaki, M. Kaldrimidou & C. Sakonidis (Eds.), *Proceedings of the 33rd conference of the International Group for the Psychology of Mathematics Education* (Vol. 2, pp. 88–96). Thessaloniki, Greece: PME.
2. Bogomolny, A. (2015). *Pythagorean Theorem and its many proofs from interactive mathematics miscellany and puzzles*. Retrieved from http://www.cut-the-knot.org/pythagoras/index.shtml
3. Boyer, C. B., & Merzbach, U. C. (1991). *A history of mathematics*. New York, NY: John Wiley & Sons.
4. Coolidge, J. L. (1968). *A history of the conic sections and quadric surfaces*. New York, NY: Dover.
5. Klein, F. (2004). *Elementary mathematics from an advanced standpoint: Geometry*. New York, NY: Dover.
6. Leung, A. (2014). Principles of acquiring invariant in mathematics task design: A dynamic geometry example. In: P. Liljedahl, S. Oesterle, C. Nicol & D. Allan (Eds.), *Proceedings of the joint meeting of PME 38 and PME-NA 36* (Vol. 4, pp. 89–96). Vancouver, Canada: PME.
7. Libeskind, S., Stupel, M., and Oxman, V. (2018). The Concept of Invariance in school Mathematics. *International Journal of Mathematical Education in Science and Technology, 49*(1), 107–120.
8. Lloyd, G. E. R. (1968). *Aristotle: The growth and structure of his thought*. Cambridge, UK: Cambridge University Press.
9. Mason, J. (1996). Expressing generalization and roots of algebra. In N. Bednarz, C. Kieran & L. Lee (Eds.), *Approaches to algebra: Perspectives for research and teaching* (pp. 65–86). Dordrecht, the Netherlands: Kluwer.
10. Sinitsky, I., and Ilany, B. (2016). *Change and Invariance. A textbook on Algebraic Insight into Numbers and Shapes*. Rotterdam / Boston / Taipei: Sense Publishers.
11. Struik, D. J. (1987). *A concise history of mathematics* (4th ed.). New York, NY: Dover Publications.

https://doi.org/10.1142/9789811218750_0010

Chapter 10

Assessing Creativity and its Development Using a Psychometric Model: The Case of Problem Posing

Atara Shriki

Introduction

As illustrated in previous chapters (Chapter 1 and Chapter 2 in this book), creativity is generally perceived as a key driver of social and economic changes. One might, therefore, anticipate that the education system would strive to nurture students' creativity. However, although teachers believe that mathematical creativity can and should be nurtured at school, the education system scarcely encourages it. As a result, most students are deprived of the opportunity to experience creative learning and thinking. As noted by Shriki (2013), this reality is a consequence of a combination of several circumstances, among them:

(i) The heavy load of teaching, the external pressure on teachers to cover the mandatory curriculum, and the expectations from teachers to make their students succeed in standardized tests. The education system puts an emphasis on achievement, and too often teachers are judged solely on the basis of their students' success in the tests. Generally, in standardized tests, learners are required to demonstrate algorithmic thinking and rote implementation of rules and are not supposed to assess students' mathematical creativity. Consequently, teachers tend to focus on memorization and algorithms, rather than on nurturing students' creativity;

(ii) Teachers are inclined to teach in the same way they themselves were taught at school (Hall, Fisher, Musanti, & Halquist, 2006). As school students as well as during their training period, most teachers did not experience mathematics education that placed an emphasis on nurturing creativity. Therefore, in the absence of a role model to draw upon, it is reasonable to assume that they may not have acquired sufficient instruments for nurturing their students' mathematical creativity;

(iii) There is a shortage of appropriate learning and teaching materials specifically designed to nurture mathematical creativity. This renders it difficult for teachers to initiate suitable learning environments for satisfying this need;

(iv) Many teachers mistakenly believe that expressions of creativity are restricted to fields, such as painting, sculpture, writing, and acting. Hence, they consider mathematics as a discipline that does not provide much room for expressing creativity (Andiliou & Murphy, 2010). Furthermore, many tend to attribute creativity to giftedness, and, therefore, refrain from taking steps to nurture the creativity of all students (Aljughaiman & Reynolds, 2005; Sriraman, 2005);

(v) There are no readily available instruments for assessing students' creativity and its long-term development. Consequently, teachers find it difficult to implement a systematic approach to develop students' creativity and monitor their progress (Shriki, 2010). Such a follow-up could provide teachers with feedback on students' progress, helping them to choose courses of action that are suitable for their students.

The main goal of this chapter, then, is to address the need for a simple instrument aimed at assessing students' mathematical creativity and its development, as well as suggest a suitable model designed for this purpose. Two main premises underpinned the development of the instrument: (a) Nurturing students' creativity is possible at all ages and ability levels and should be an integral part of the mandatory curriculum; (b) Providing teachers with a practical and straightforward instrument for assessing students' creativity and its progress might inspire teachers to incorporate appropriate activities in their daily teaching.

The described model is designed to assess mathematical creativity in the context of problem posing. The model is based on Torrance's (1974) psychometric approach to measuring creativity and uses four indices: fluency, flexibility, originality, and organization or elaboration (see also Chapter 1). Relating to creativity in terms of fluency, flexibility, originality, and organization provides teachers with an easy schema for its assessment (Brandau & Dossey, 1979; Silver, 1997).

Description of the Model

As stated, the model for assessing creativity in the context of problem posing uses four indices: fluency, flexibility, originality, and organization. Drawing on Balka (1974), Brandau and Dossey (1979), and Torrance (1974), fluency is measured by the number of different problems posed; flexibility is measured by the number of different categories of the posed problems; originality is measured by the relative infrequency of the problems; and organization is measured by the number of problems formulated as generalizations. As pointed out by Brandau and Dossey (1979), all of these variables are highly significant and positively correlated.

The view of creativity as a personal trait that can be developed in school students, requires a distinction between relative and absolute creativity (Leikin, 2009; Sriraman, 2005). While absolute creativity is associated with remarkable historical works of prominent mathematicians, relative creativity refers to discoveries made by a specific person within a specific reference group. This distinction is expressed in Leikin's (2009) model, designed to assess mathematical creativity in the context of multiple solutions to a given problem (see Chapter 2). Adapting this distinction to the context of problem posing, in the current model, each student receives a total as well as relative score for fluency, flexibility, originality, and organization, as will be explained below.

Guidelines for scoring the four indices of creativity

The scoring process refers to the third level of the "What-if-Not?" (WIN) strategy (see Chapter 7), whereby learners have to relate to the list of attributes they have produced (first level) and the possible alternatives of the attributes (second level) for posing new problems. In order to determine students' score for each aspect of creativity, as well as an overall creativity score, the following steps are applied (Shriki, 2013):

Fluency scoring

The student's total score for fluency is determined by counting the number of different new problems he or she posed, based on a given problem. In other words, 1 point for each problem that was posed. The relative score for fluency is determined in the following manner: the student in the reference group who received the highest total score for fluency is given a score of 100 for relative fluency. All other students' relative scores for fluency are determined according to the highest score. For example, if the highest total score is 20 (namely, each student posed 20 different problems at the most), then those who posed 20 problems receive a relative score of 100. Those who posed, for instance, 12 problems, receive a relative score of 60.

Flexibility scoring

Student's total score for flexibility is determined by the overall number of different categories that are constituted by the posed problems. The relative score for flexibility is determined similarly to the relative score computation used for fluency.

Originality scoring

Since originality is, by its very nature, relative, it is necessary to predetermine the condition for originality. If we take, for example, all the problems posed by one-third of the students as the "upper" limit for originality, then all students who posed problems that had been posed by 33% or less of the students, will receive a score for originality. Obviously, this "upper limit" can vary from one class to another. Other students will receive a score of 0. The relative score for originality is determined in the following manner: Assume that the group consists of 30 students. In this case, only problems that were posed by 10 or fewer students are considered for scoring originality. The students who posed the largest number of such problems receive a score of 100 for originality. Other students who have posed problems that are considered for scoring originality, receive a relative score similarly to the scoring fluency and flexibility.

Organization (or generalization) scoring

Students' total score for organization is determined according to the number of posed problems that are formulated as a generalization. Then, the relative score is computed similarly to the above cases.

An overall total score for creativity

Finally, the overall score for creativity is determined by assigning a weight for each of the four relative scores. Deciding on the relative weighting is subjected to the teachers' preferences, according to the importance they ascribe to each aspect. Clearly, the beliefs teachers hold about creativity are likely to affect the way they define, operationalize, and assess students' creativity (Andiliou & Murphy, 2010). Hence, it is important to grant teachers the freedom to determine the weight of each component, adapting it to their choices, emphases and teaching goals.

Demonstrating the Scoring Process

In this section, I demonstrate the scoring process, illustrating how to set the total and relative scoring for fluency, flexibility, and organization; the scoring for originality; and the final overall scoring for creativity.

This demonstration is based on results taken from a study conducted by Lavy & Shriki (2010) and fully presented in Shriki (2013). In this study, 25 pre-service teachers engaged in geometric problem posing by implementing the WIN strategy. The given problem was taken from Hönsberger (1985, p. 81):

Triangle *ABC* is inscribed in circle *O*. *D* is a variable point on the circumference of *O*. Perpendiculars are drawn from *D* to *AB* and *AC*. *E* and *F* are intersection points of the perpendiculars with the side of the triangle, respectively. Determine the position of *D* whereby *EF* is of maximal length (Figure 1).

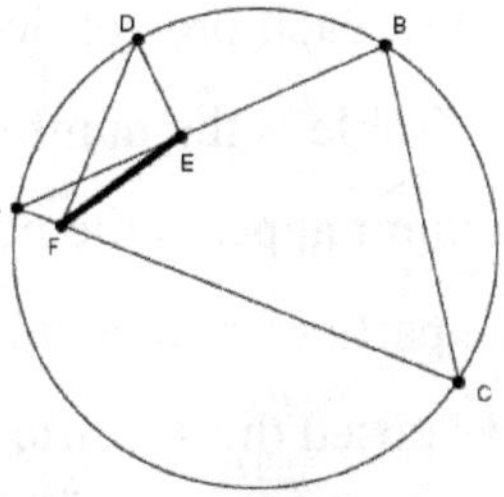

Figure 1:
The given problem

Based on this problem, the pre-service teachers posed 46 different problems that were classified into 12 different categories. Each problem was assigned a number indicating its category. For example, category No. 1 was “A polygon is inscribed in a circle”. Namely, the only change related to the type of polygon inscribed in the circle. Four different alternative problems related to this category. These new problems were numbered accordingly: 1.1 A quadrangle is inscribed in the circle; 1.2 A square is inscribed in the circle; 1.3 A pentagon is inscribed in the circle; and 1.4 an n-sided polygon is inscribed in the circle. Problems 1.1, 1.2, 1.3, and 1.4 were posed by 25 pre-service teachers (100%), Twenty pre-service teachers (20%), 14 pre-service teachers (56%), and 4 (16%) pre-service teachers, respectively.

Arranging the Data

In order to “map” the problems that were posed by the pre-service teachers’, Table 1 was generated. In the first line appears the number (1-25) of each pre-service teacher (PT). The left column in the grey part of the Table is the number of each category (c.), 1-12, and in the second left column appears the number assigned for each problem that was posed in the particular category. For each PT, a mark of “+” indicates that he or she has posed the specific problem.

Scoring fluency, flexibility, originality, organization, and creativity

Scoring total and relative fluency

For each PT, the number of problems he or she posed was counted, and this determined his or her total score for fluency (see Table 1- Flu, T.S). For example, PT1 posed 16 problems, thus PT1's total score for fluency was 16. The relative score (R.S) for fluency was determined in the following manner: The PTs with the highest number of problems posed received a relative score (R.S.) of 100 for fluency. Table 1 indicates that PT10 has posed more problems than the other PTs (30 problems, T.S=30), and thus received a relative score of 100. Other scores were determined relatively to PT10. For example, PT1, who posed 16 problems, received a R.S. of 53 [round of $\frac{16}{30} \cdot 100$]. Other R.Ss for fluency were calculated in a similar manner.

Scoring total and relative flexibility

For each PT, the number of categories he or she posed problems was used for determining the total score for flexibility. For example, PT1 posed problems that corresponded to 7 of the 12 categories, thus her total score for flexibility was 7.

The PTs with the highest number of categories to which their problems corresponded received a relative score of 100 for flexibility. Table 1 indicates that PT16 posed problems that matched all the 12 categories that were generated, and thus received a relative score of 100. Other R.Ss were determined relatively to PT16. For example, PT1's problems matched 7 categories, thus her relative score for flexibility was 58 [round of $\frac{7}{12} \cdot 100$]. Other R.Ss for flexibility were calculated in a similar manner.

Scoring total and relative originality

We shall now examine the two last right columns of Table 1, entitled "Orig". For each problem that was posed, the number of PTs who posed it was counted and recorded in the penultimate right column (entitled "Tot"). For example, problem 1.1 was posed by all 25 PTs. The problem with the highest number of PTs who suggested it was indicated by "100%" (see the ultimate right column of Table 1). Other problems were calculated relatively to problem 1.1 (or other problems that were posed by all the PTs). For example, problem 12.1 was posed by six PTs, and thus was indicated by 24% [$\frac{6}{25} \cdot 100$].

As mentioned, for scoring originality, an arbitrary decision was made to include only problems posed by a third or less of the pre-service teachers. Namely, problems recorded in the ultimate right column as "33%" or less. In order to determine each PT's total score for originality, the number of problems posed that were indicated by "33%" or less were counted. This constituted the total score for PTs' originality (see Table 1-Orig, T.S). For example, PT1 did not pose any such problems, thus her T.S for originality was 0.

The pre-service teachers with the highest number of such problems received a relative score of 100 for originality. Table 1 illustrates that PT10 posed 7 problems that were suggested by less than 33% of the PTs. Her total score for originality was 7 and her relative score for originality was 100. Other scores for relative originality were determined based on PT10's score. For example, PT12 suggested 5 such problems, thus her score for originality was 71 [round of $\frac{5}{7} \cdot 100$]. Other scores for relative originality were calculated in a similar manner.

Scoring total and relative organization

The score for organization was assigned in accordance with the number of problems that were posed as generalizations. There were only five such problems (1.4, 5.6, 5.10, 8.2, and 9.2). Table 1 indicates that two PTs (PT12 and PT13) posed two generalized problems. Therefore, their total score for organization was 2 (Org, T.S) and their relative score for organization was 100. Those who posed one generalized problem received a total score of 1, and a relative score of 50. The remainder received a total and relative score of 0.

Scoring overall creativity

Finally, in order to assign a score for each pre-service teacher's overall creativity, it is necessary to determine the relative weight of fluency, flexibility, originality, and organization. As mentioned, the relative weight should reflect the importance teachers ascribe to each component, as well as their priorities, preferences, emphases given in class, and teaching goals. In this example, merely for a matter of simplicity, each component was arbitrarily assigned the same weight (25%), namely, a round of the sum of all R.Ss divided by 4, as can be seen from the last row of Table 1.

The following are some comments relating to the implementation of the model and its implications:

(i) Adequacy of the posed problems: the appropriateness relates, among others, to the appropriate formulation of the problem and to the completeness of the information (namely, there is no lack of information or information that is not relevant) (Van Harpen & Presmeg, 2013; Van Harpen & Sriraman, 2013). Posing mathematical problems cannot be an end in itself. In other words, asking students to pose mathematical problems is not enough, and it is essential to instruct them explicitly to examine the appropriateness of the problems they pose (Andiliou & Murphy, 2010). The WIN strategy does provide a great deal of freedom in posing problems. However, given that students receive scores for fluency, they may be tempted to pose as many problems as possible without checking their appropriateness.

Table 1
Pre-service teachers' relative and total scores for fluency, flexibility, originality, organization and overall creativity.
(PT = pre-service teacher, c = category, Flue = fluency, Flex = flexibility, Orig = originality, Org = organization, T.S = total score, R.S = relative score).

C.	PT	1	2	3	...	10	11	12	13	14	15	16	17	...	23	24	25	Orig. Tot	Orig. %
1	1.1	+	+	+	...	+	+	+	+	+	+	+	+	...	+	+	+	25	100
	1.2	+	+	+	...	+	+	+	+	+	+	+	+	...				20	80
	1.3				...			+	+	+	+	+	+	...	+	+	+	14	56
	1.4				...			+	+	+	+			...				4	16
2	2.1	+	+	+	...	+	+	+	+	+	+	+	+	...	+	+	+	25	100
	2.2	+	+	+	...	+	+	+	+	+	+	+	+	...	+	+	+	25	100
⋮	⋮	⋮	⋮	⋮	⋮	⋮	⋮	⋮	⋮	⋮	⋮	⋮	⋮	⋮	⋮	⋮	⋮	⋮	⋮
12	12.1				...							+	+	...				6	24
	12.2				...									...			+	1	4
Flue	T.S	16	17	17	...	30	29	29	29	29	29	28	25	...	19	16	17		
	R.S	53	57	57	...	100	97	97	97	97	97	93	83	...	63	53	57		
Flex	T.S	7	7	7	...	9	9	9	9	9	11	12	10	...	8	8	9		
	R.S	58	58	58	...	75	75	75	75	75	92	100	83	...	67	67	75		
Orig	T.S	0	0	0	...	7	6	5	5	5	5	5	2	...	3	2	3		
	R.S	0	0	0	...	100	86	71	71	71	71	71	29	...	43	29	43		
Org	T.S	0	0	0	...	1	1	2	2	1	1	0	0	...	1	0	0		
	R.S	0	0	0	...	50	50	100	100	50	50	0	0	...	50	0	0		
Creativity		28	29	29	...	81	77	86	86	73	77	66	49	...	56	37	44		

(ii) Refraining from placing too much emphasis on the need to solve the problems posed: As noted in Chapter 7, the implementation of the WIN strategy may lead to stimulating inquiry problems. However, instructing students to solve the problems they posed may reduce their willingness to pose "wild" problems, fearing that they will not be able to solve the problems. Therefore, in order not to restrict students to posing trivial problems, thereby limiting their creativity, teachers are advised to instruct their students not to worry too much about the need to solve all the problems they pose. "Difficult" problems can be solved in class in a collaborative manner (Lavy & Shriki, 2010).

(iii) Class discussion: Interaction between learners enables them to exchange inspiring ideas and support the development of their creativity. Indeed, it appears that much of the breakthroughs of creative people occurred as a result of collaboration with others (John-Steiner, 2000). It is, therefore, essential to provide students with opportunities to share knowledge and encourage creative students to communicate their ideas and insights with their peers (Sriraman, 2005). Since among the four psychometric indices of creativity, originality is more difficult to develop than the others, it is suggested to pay special attention to this component. For the purpose of fostering students' ability to appreciate the meaning of originality, it is important to present them with examples of original problems posed by their classmates, discussing issues related to the nature of these problems. This will allow students to learn how to distinguish between "ordinary problems" and "original problems" and to assess the degree of originality of the problems they pose.

Assessing the Development of Creativity

One of the reasons for teachers' reluctance to assess students' creativity is the feeling that they are unable to follow its development as a response to activities in which their students engage (Shriki, 2010). Teachers regularly monitor the progress of their students through numerical information, such as scores, which provides them with feedback on their teaching and the need to generate changes (Wiliam, Lee, Harrison, & Black, 2004). Consequently, the question is how the use of the model over time in learning environments designed to nurture students' creativity, may enable teachers to monitor the development of creativity. The following are several suggestions on how the resulting numerical data can be used.

Graphical display of data

Graphical display of data can include various types of information. For example, the relative scores of every student for each of the four indices of creativity, as well as the score for creativity (see Figure 2); distribution of total and relative scores (see Figure 3); and more. Graphical display can also portray changes in the relative scores of each student, for each index, over time.

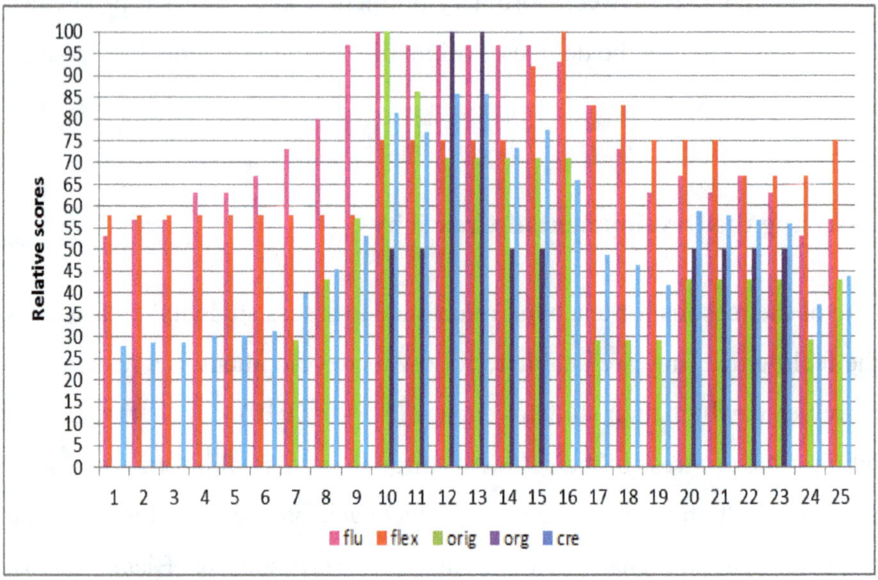

Figure 3: Relative scores of every pre-service teacher (1-25)

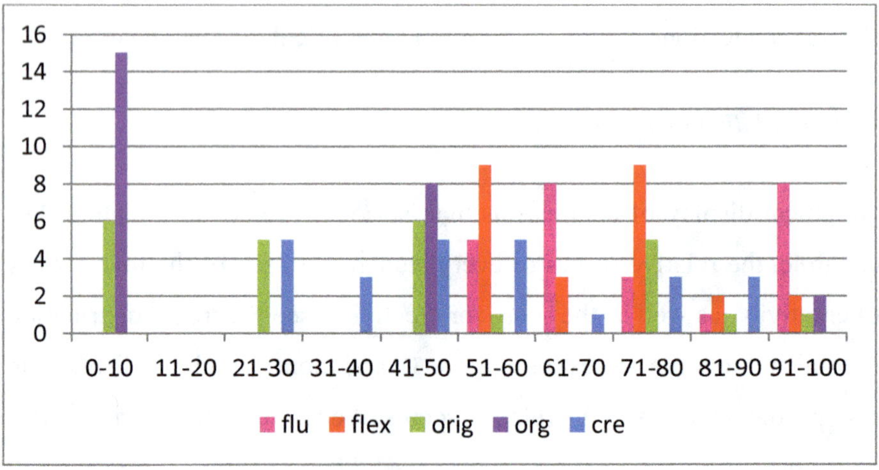

Figure 4: Distribution of relative scores

Based on the various graphical displays, teachers can get an idea about the strengths and weaknesses of each student as well as of the entire class. Thus, they can make some pedagogical decisions regarding the emphases they should put, in order to nurture students' creativity.

Evaluating the development of creativity

Assuming that teachers engage their students in problem posing assignments for an extended period of time, how can they employ the model for assessing the *development* of their students' creativity? What indicators might serve this purpose?

The model allows producing the following four numerical values: total scores, relative scores, mean values, and standard deviations. The strengths and weaknesses of these numerical values for making educational decisions are as follows:

Total scores

The total scores for fluency, flexibility, originality, and organization cannot be applied for assessing the development of creativity. This is due to the fact that the number of problems that can be posed by employing the WIN strategy, and the resulting categories, might depend on the richness of attributes that are embedded in the original problem. However, total scores might be used in cases whereby teachers wish to make comparisons between classes that work on the same problems. Such a comparison is valuable mainly when teachers wish to examine the impact of different approaches designed to engage their students in problem posing assignments. Moreover, the comparison is important when they want to compare the effect of the same approach (e.g., the WIN strategy) on students studying mathematics at different levels or in different age groups.

Mean relative scores

In the case of relative scores, a substantial increase in the mean value of any of the relative scores indicates that some students developed their creativity much more than others. In such cases, special attention should be paid to students whose relative scores did not change much with time or even decreased. Alternately, this may support theories claiming that creativity is innate (e.g., Sternberg & Lubart, 1991) and that students who are more creative from the outset show a greater tendency to keep developing their creativity. Minor or no changes might imply that the entire class, as a reference group, either did not progress or all students exhibited similar progress (or regression).

Standard deviations

Assuming that practicing the WIN strategy over and over again will not entail a decrease in mathematical creativity, a gradual decrease of standard deviation in any of the measures implies the development of creativity of the class as a whole and convergence towards demonstrating a similar degree of creativity. Obviously, this is the most desirable situation. An increase or decrease in standard deviations can, therefore, serve as a primary indicator for examining the development, lack of development, or perhaps a decrease of creativity attributed to the class as a group of reference. If indeed, the creativity of all students can be developed through appropriate education (e.g., Fryer, 1996; Henry, 2009), then we would expect a gradual reduction in standard deviations over time. However, a gradual increase in standard deviation might support theories advocating that creativity is innate and that only exceptional people can demonstrate creative behavior. It should be noted that in order to be able to track relative changes, it is necessary to predetermine the "upper limit" used for scoring originality, as well as the weight of each component of creativity, and adhere to it. Changing the upper limit or relative weight, though, might be pedagogically justified, preventing the possibility of tracking the development of creativity.

Self-Assessment of the Development of Creativity

The standards of the National Council of Mathematics Teachers in the United States (NCTM, 2000) recommend engaging students in self-assessment activities in order to develop their confidence as learners of mathematics and to develop their self-regulated learning skills for the benefit of their future lives. These skills will help them attain the goals they set for themselves as well as feel ownership of and responsibility for their learning. This will also provide students with the opportunity to reflect on their learning and increase their motivation to study (Smith, 1997), and to refine their products in successive iterations (Kaufman, Plucker, & Russell, 2012). Moreover, students' involvement in assessing their own progress in learning, adds reflective and metacognitive dimensions to their learning. Students enjoy participating in self-assessment and observing their progress by using graphical displays and are able to use such displays for articulating the value of their own studies (Brookhart, Andolina, Meganzuza, & Furman, 2004). However, in view of helping students to assess their performance and progress, they should be provided with clear guidelines that are easy to follow (Enz & Serafini, 1995), as well as support them while they implement those assessment guidelines (Brookhart et al., 2004). Research indicates (e.g. Enz & Serafini, 1995) that after gaining suitable training, students are able to keep the self-assessment with little or no assistance from their teachers. Furthermore, students who are trained in self-assessment, outperform their peers who have not received such preparation. Relating specifically to self-assessment of one's own creativity, Chamberlin & Moon (2005) believe that such process in itself requires creativity and, in return, will enable students to refine their products in successive iterations. Therefore, it is recommended to enable students to assess their own creativity and its development.

When students are provided with a graphical representation of their relative progress in each one of the indices of creativity, they receive feedback that allows them to self-assess their development. Reflecting on such information enables students to identify their strengths and weaknesses, both relative to their previous attainments and to their reference group (Shriki, 2013). Figure 5 illustrates a possible personal graphical display of cumulative scores for each of the indices of creativity for five WIN tasks (t1-t5).

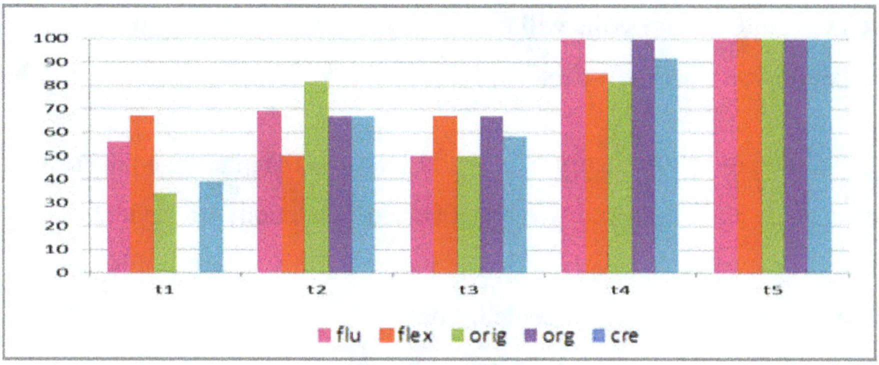

Figure 5: An example of a personal graphical display of cumulative relative scores (t=task)

It should be noted, however, that self-assessment of creativity may not be beneficial for all students. As could be seen in Chapter 1, there are various approaches to the study and assessment of creativity. One such approach is the Investment Theory that is rooted in the social-personality approach of Sternberg & Lubart (1995). The theory seeks to understand the foundation of creativity and to explain why some people are more creative than others. According to this theory, performance results from a confluence of six interrelated resources: intelligence, knowledge, intellectual styles, personality, motivation, and environment. The first five resources are person-centered, while the sixth one is context-centered. "*In order for these resources to be used effectively, they must converge in a way that capitalizes upon them both singly and in interaction*" (Sternberg & Lubart, 1995, p. 4). In some cases, 'strong' resources may compensate for 'weak' resources. For example, a high level of motivational resource may counteract a poor environmental context. However, a high level of relevant intellectual skills will not fully compensate for the unwillingness to take risks (which is a personality trait).

Drawing on this theory, Shriki and Lavy (2014) interpreted the results of a study that followed the benefits and limitations of students' self-assessment of their creativity, using the model discussed in this chapter. Two case studies were presented, both of 11th-grade students, Ruth and Michael. These two students represent two types of learners. While Ruth's relative scores consistently increased, Michael's relative scores remained almost unchanged. Ruth and Michael's mean scores in mathematics were 86 and 82, respectively. Figure 6 presents Ruth and Michael's cumulative relative scores for fluency, flexibility, originality, organization, and creativity for the five tasks (t1-t5). The score for creativity was calculated so that each of the four aspects was given an equal weight.

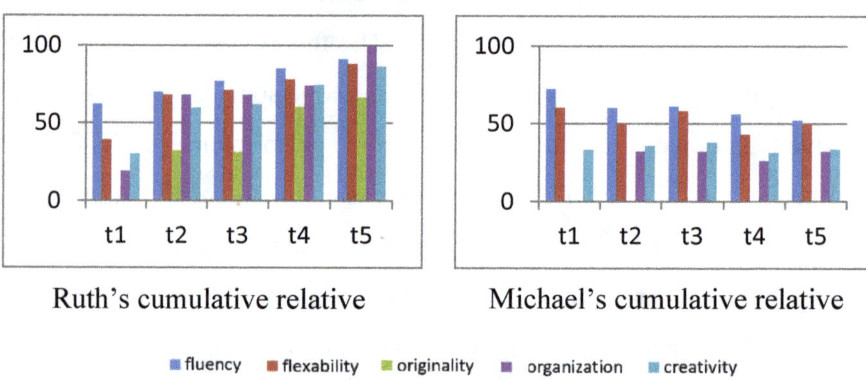

Ruth's cumulative relative Michael's cumulative relative

Figure 6: Ruth and Michael's cumulative scores for task1 (t1) to task5 (t5)

Figure 6 clearly indicates that, unlike Chamberlin & Moon's (2005) observation, self-assessment of creativity has a diverse effect on different students. Therefore, the question that was put to the test by Shriki and Lavy (2014) related to the factors, combined with the self-assessment process, that affect the development of students' creativity. To answer the question, the researchers employed the Investment Theory for analyzing Ruth and Michael's reflections on the process. The results indicated that self-assessment of creativity can be beneficial for students who possess an optimal mixture of resources from the outset. Such students are able to exploit this process for developing it further. On the other hand, students who lack a certain degree of threshold for some resources might be harmed by the process and their creativity might be suppressed. In other words, self-assessment of creativity based on this model and the graphical display that can be derived from it, are not equally suitable for all students.

Summary

In order to respond teachers' need for a coherent and accessible instrument that will enable them to monitor and assess the development of students' creativity, this chapter presents a dedicated model intended to assess creativity and its development based on the psychometric approach. The model uses four indices: fluency, flexibility, originality, and organization. It is particularly suitable for assessing the degree of creativity in a learning environment based on the WIN strategy (or another approach to problem posing). It should be noted, however, that although teachers are used to assessing students' learning outcomes by means of numerical scores, the complexity of the issues associated with creativity and its assessment requires the consideration of aspects related to learners' personal resources as well.

References

1. Aljughaiman, A., & Reynolds, E. (2005). Teachers' conceptions of creativity and creative students. *Journal of Creative Behavior, 39*(1), 17–34.
2. Andiliou, A., & Murphy, K. P. (2010). Examining variations among researchers and teachers' conceptualizations of creativity: A review and synthesis of contemporary research. *Educational Research Review, 5*, 201–219.
3. Balka, D. S. (1974). Creative ability in mathematics. *Arithmetic Teacher, 21*(7), 633–363.
4. Brandau, L. I., & Dossey, J. A. (1979). *Processes involved in mathematical divergent problem-solving.* San Francisco: American Educational Research Association.
5. Brookhart, S., Andolina, M., Zuza, M., & Furman, R. (2004). Minute math: An action research study of student self-assessment. *Educational Studies in Mathematics, 57*(2), 213–227.
6. Chamberlin, S. A., & Moon, S. (2005). Model-eliciting activities: An introduction to gifted education. *Journal of Secondary Gifted Education, 17*, 37–47.
7. Enz, B., & Serafini, F. (1995). Involving students in the assessment process. *Teaching PreK-8, 25*(5), 96–97.
8. Fryer, M. (1996). *Creative teaching and learning*. London: Paul Chapman Publishing Ltd.
9. Hall, L. D., Fisher, C., Musanti, S., & Halquist, D. (2006). Professional development in teacher education: What can we learn from PT3? *Tech Trends, 50*(3), 25–31.
10. Henry, J. (2009). Enhancing creativity with M.U.S.I.C. *The Alberta Journal of Educational Research, 5*(2), 199–211.
11. Hönsberger, R. (1985). *Mathematical gems III.* The Mathematical Association of America. The United States of America.

12. John-Steiner, V. (2000). *Creative collaboration*. New York, NY: Oxford University Press.
13. Kaufman, J. C., Plucker, J. A., & Russell, C. M. (2012). Identifying and assessing creativity as a component of giftedness. *Journal of Psychoeducational Assessment 30*(1), 60–73.
14. Lavy, I., & Shriki, A. (2010). Engaging in problem-posing activities in a dynamic geometry setting and the development of prospective teachers' mathematical knowledge. *Journal of Mathematical Behavior, 29,* 11–24.
15. Leikin, R. (2009). Exploring mathematical creativity using multiple solution tasks. In: R. Leikin, A. Berman, & B. Koichu (Eds.), *Creativity in mathematics and the education of gifted students* (pp. 129–145). Rotterdam, the Netherlands: Sense Publisher.
16. NCTM - National Council of Teachers of Mathematics (2000). *Principles and standards for school mathematics*. Reston, VA: NCTM.
17. Shriki, A. (2010). Working like real mathematicians: Developing prospective teachers' awareness of mathematical creativity through generating new concepts. *Educational Studies in Mathematics, 73*, 159–179.
18. Shriki, A. (2013). A model for assessing the development of students' creativity in the context of problem posing. Creative Education, 4(7), 430–439.
19. Shriki, A., & Lavy, I. (2014). Students' self-assessment of creativity: Benefits and limitations. In Nicol, C., Oesterle, S., Liljedahl, P., & Allan, D. (Eds.), *Proceedings of the Joint Meeting of PME 38 and PME-NA 36, 5,* 177–184. Vancouver, Canada: PME.
20. Silver, E. A. (1997). Fostering creativity through instruction rich in mathematical problem solving and problem posing. *ZDM - The International Journal on Mathematics Education, 29*(3), 75–80.
21. Smith, C. (1997). Student self-assessment at St. Bernadette's primary school. *Primary Educator, 3*(4), 7–9.
22. Sriraman, B. (2005). Are giftedness and creativity synonyms in mathematics? *The Journal of Secondary Gifted Education, 17*(1), 20–36.
23. Sternberg, R. J., & Lubart, T. I. (1995). *Defying the crowd: Cultivating creativity in a culture of conformity*. New York, NY: Free Press.

24. Torrance, E. P. (1974). *The Torrance tests of creative thinking: Technical-norms manual*. Bensenville, IL: Scholastic Testing Services.
25. Van Harpen, X. Y., & Presmeg, N. C. (2013). An investigation of relationships between students' mathematical problem-posing abilities and their mathematical content knowledge. *Educational Studies in Mathematics, 83*(1), 117–132.
26. Van Harpen, X.Y., & Sriraman, B. (2013). Creativity and mathematical problem posing: An analysis of high school students' mathematical problem posing in China and the USA. *Educational Studies in Mathematics, 82*(2), 201–221.
27. Wiliam, D., Lee, C., Harrison, C., & Black, P. (2004). Teachers developing assessment for learning: Impact on student achievement. *Assessment in Education: Principles, Policy & Practice, 11*(1), 49–65.

Epilog

Photo by S. Kirsner

https://doi.org/10.1142/9789811218750_bmatter

Creativity is just connecting things. When you ask creative people how they did something, they feel a little guilty because they didn't really do it, they just saw something. It seemed obvious to them after a while. That's because they were able to connect experiences they've had and synthesize new things.

Steve Jobs, Apple Inc.

At the end of this book, we challenge the readers with seven geometric problems, that call for implementing divergent and non-algorithmic thinking, or, in Steve Jobs' words, to connect experiences and synthesize new things.

We wish the readers pleasure in dealing with the problems.

1. A five-pointed star

What is the sum of the angles in a five-pointed star?

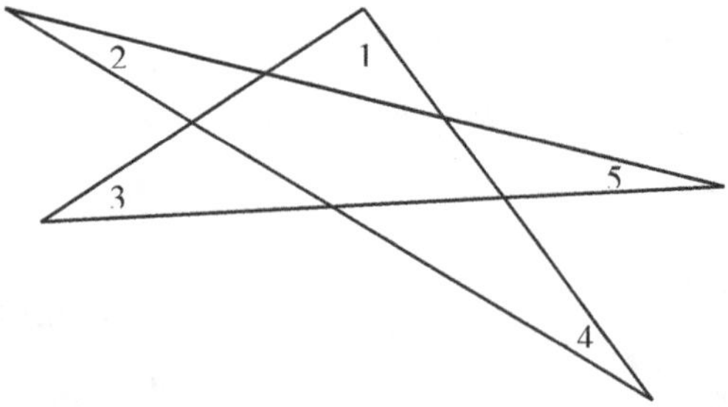

Can you find a generalization to the sum of the angles in any star with an odd number of sides?

2. Good neighborhood[1]

Eda and Azusa each own a piece of land that lies between the same pair of lines.

Their common boundary is formed by a bent line segment as shown.

The problem is to change the bent line into a straight-line segment that still divides the region into two pieces, each with the same area as before.

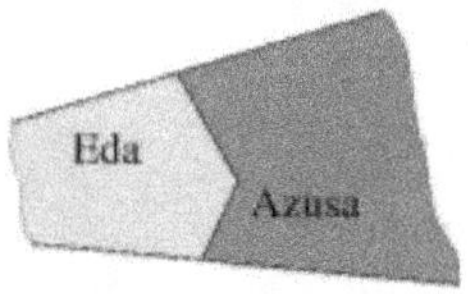

---

[1]Taken from Siegal, A. (2001). Telling lessons from the TIMSS videotape: Remarkable teaching practices as recorded from eighth-grade mathematics classes in Japan, Germany, and the U.S. (p. 5).
Available at: https://cs.nyu.edu/faculty/siegel/ST11.pdf

3. Drawing medians in a triangle

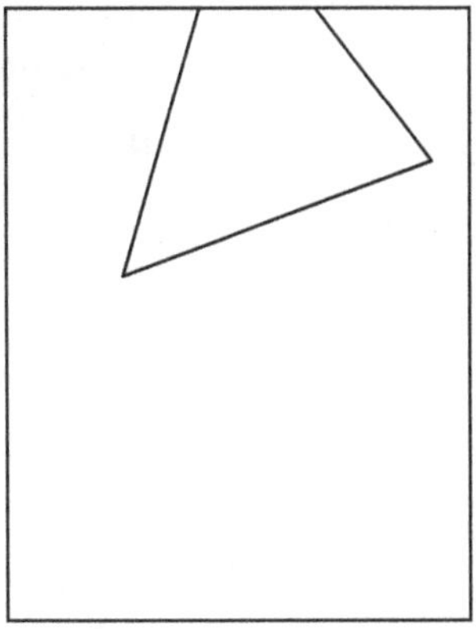

The mathematics teacher asked the pupils to draw a triangle and find the intersection point of its medians.

Ruth started drawing a triangle. However, she soon found out that she had not planned the drawing well and that she would not be able to complete the triangle.

This is the notebook page on which Ruth has drawn the triangle.

Can Ruth somehow find the meeting point of the medians?

4. A rhombus-shaped swimming pool[2]

At the center of the city, there is a circle-shaped park. The city council wants to build in it a rhombus-shaped swimming pool. The Mayor invited an architect in order to execute this project.

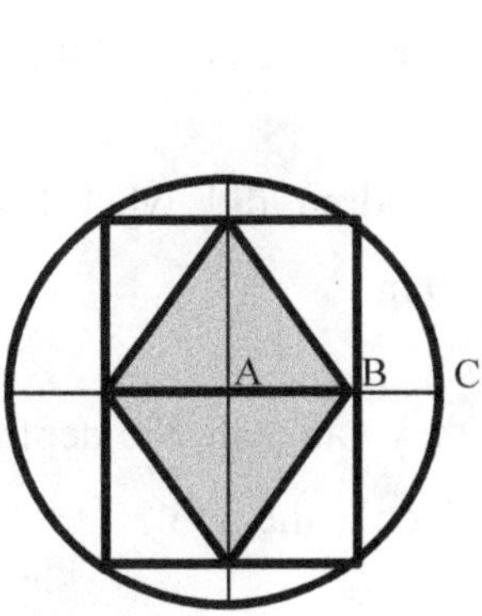

The Mayor said: "I am interested in building a rhombus-shaped pool in the following way but what is the length of each side of the pool?"

Can we find the length of the pool side if: AB = 20m; BC = 15m?

---

[2]Based on the problem "Dimensions in Disguise", in Gardner, M. (1999). *Aha! Insight*. Scientific American, Inc/W.H. Freeman and Company, New York City (p. 34).

5. Water lily in the pond[3]

A water lily was growing in a pond. At the peak of its bloom, it reached the height of 10 cm above water level.

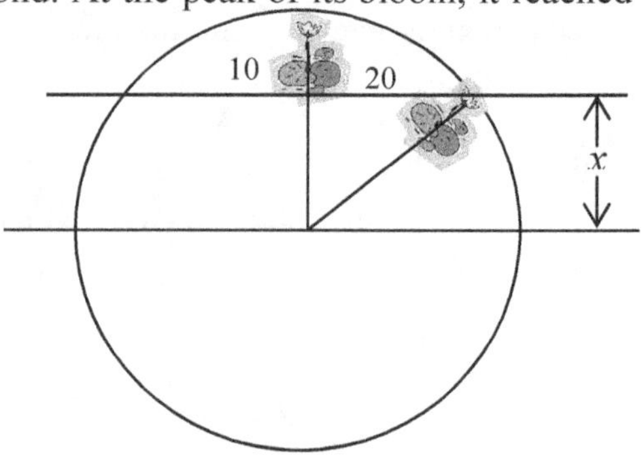

Children who came to the pond wondered how deep was the water in the pond.

Ruth suggested: "One of us will place a finger at the place where the stem touches the water. Then we will rotate the flower until its head touches the water and measure the distance between the place of the finger and the place which the head of the flower reached".

The children did what Ruth suggested and found that the distance was 20 m.

Can they calculate the depth of the pool? If no – please explain. If yes – how deep is the pool?

---

[3]Based on the elaboration to the problem "Dimensions in Disguise", in Gardner, M. (1999). *Aha! Insight*. Scientific American, Inc/W.H. Freeman and Company, New York City (p. 35).

6. Dana's golden ring[4]

On the occasion of her birthday, Dana ordered a golden ring in a jewelry shop. She gave the jeweler the following drawing, indicating that the length of the segment AB, that is tangent to the inner circle, is 3 cm.

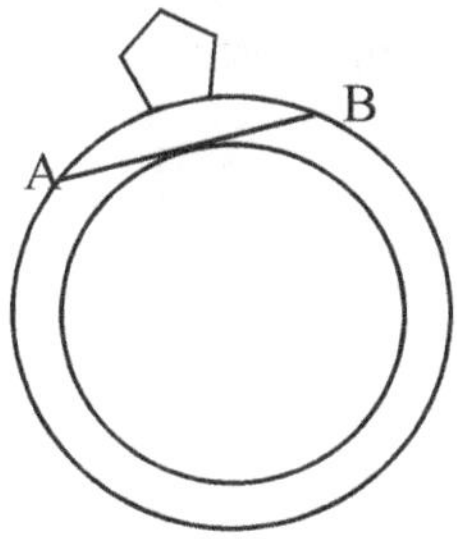

The jeweler told Dana that the price of the ring would be determined according to the area between the two circles. He asked her to return the next day in order to hear the quotation.

After Dana left, the jeweler looked at the drawing and was surprised to realize that he could not calculate the area between the two circles.

Can we help the jeweler?

[4]Based on the problem "Carpet Confusion", in Gardner, M. (1999). *Aha! Insight*. Scientific American, Inc/W. H. Freeman and Company, New York City (p. 54).

7. Moving boats[5]

A boat is floating in the harbor, tied to a rope.

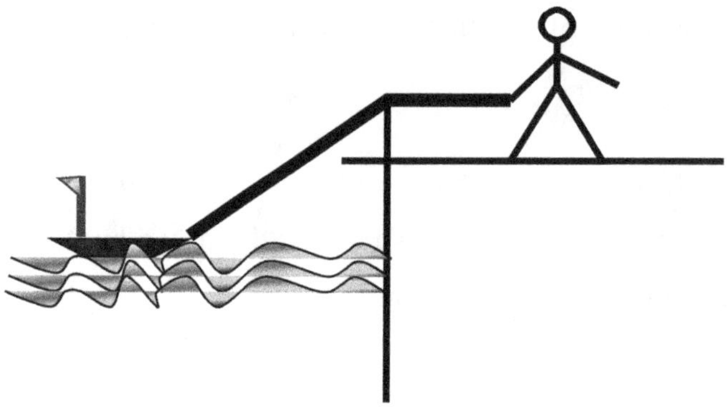

The boat owner wants to pull the boat to the pier. For that purpose, he uses a rope, pulling the boat towards the wall of the pier. When the boat owner passes one meter, what is the distance that the boat passes?

Exactly one meter / less than one meter / more than one meter.

[5]Based on the activity "The Rowbaot Paradox", in Movshovitz-Hadar, N., & Webb, J. (1998). *One equals zero and other mathematical surprises.* Key Curriculum press (p. 56).

https://doi.org/10.1142/9789811218750_bmatter

About the editors / writers

Dorit Patkin is a Full Professor (Ph.D.) in mathematics education. Currently works at the mathematics education faculty of sciences, Kibbutzim College of Education. She was the head of the mathematics department for 8 years and the coordinator at the educational television channel as well as the scientific consultant in this area. Now she is the head of the M.Ed. program in mathematics education and the M.Teach for high school. Her research interests are teacher education and teaching methods, math teaching and education, teaching and learning geometry, exploring students' mistakes and misconceptions in math and professional development of math teachers. Her current project is 'Geometry in 2-D & 3-D'.

Atara Shriki is an Associate Professor in mathematics education (Ph.D.). Currently works at the mathematics education faculty of sciences at the Kibbutzim College of Education, teaches undergraduate and graduate students, and supervises graduate and postdoctoral students. She also works at KESHER CHAM- R&D center for mathematics education at Technion — Israel Institute of Technology, and headed a Ministry of Education program designed to reduce the gap between boys and girls. Her research interests concern the professional development of mathematics teachers, the nurturing of mathematical creativity, STEAM education, the history of mathematics, and aspects related to online teaching and learning of mathematics.

Ilana Levenberg is an Associate Professor in mathematics education (Ph.D.). She is currently Head of M. Teach Program at "Gordon" - Academic College of Education in Haifa. Previously, was Head of the Mathematics department. Her background is in high school mathematics teaching. Also served as a mathematics education academic coordinator and initiated various mathematics school projects. Supervisor of student's graduate works. Her research interests include mathematics teaching, particularly geometry. Published three co-authored geometry books (with prof. Patkin) and papers about the professional development of mathematics teachers, teaching gifted students and matriculation exams.

About the writers

Dr. Ronit Bassan-Cincinatus – Head of mathematics department in the faculty of sciences at the Kibbutzim College of Education, Technology and the Arts. A senior lecturer of mathematics education, highly seasoned in-service and pre-service teachers. Among her research fields and areas of interest is the education of mathematics teachers for ages ranging between preschoolers and up to post-primary students, developing education strategies while emphasizing common mistakes and misconceptions, and deepening teachers' and students' understanding of mathematical models' application for mathematical problem solving.

Dr. Eli Appelboim holds a Ph.D. degree in Mathematics from the Technion. He is a research engineer at the Technion – IIT, faculty of Electrical Engineering, and also a lecturer at the Gordon Academic College of Education, School of Mathematics Education. His research expands over three main fields: The first field is pure mathematical research of the topology and geometry of 3-dimensional manifolds. The second field is applied mathematics research in the areas of information theory and image processing. The third direction of research is mathematics education in general with a focus on the teaching of geometry in particular.

Associate Professor Ilya Sinitsky (Ph.D.) is a mathematician who has worked for over 30 years in the field of mathematics education. He is a lecturer at Gordon Academic College of Education and at Shaanan College, Haifa, Israel. His teaching activity and research interests are focused on the development processes of students' mathematical thinking of students and on ways of integrating general mathematical principles in learning mathematics. Recently he published together with B. Ilany (the monograph "Change and Invariance: A Textbook on Algebraic Insight into Numbers and Shapes" (in Sense Publishing House).

Associate Professor Moshe Stupel is a mathematics educator at Shaanan College and at Gordon Academic College of Education, Haifa, Israel, dealing with the existing attributes of different geometric shapes. Has extensive experience in the study of dynamic properties using computerized technology. Published about 100 articles in journals of Technology and Mathematics Education and has published two books.

Furthermore, he developed mathematical enrichment materials. Previously was a principal of a six-year high school.

https://doi.org/10.1142/9789811218750_bmatter

Name index **Page**

https://doi.org/10.1142/9789811218750_bmatter

Subject index **Page**

Subject index	**Page**
generalization	57, 67, 68, 72, 85, 92, 95, 124, 126, 160, 161, 165, 169, 171, 182, 213, 272, 295, 300, 302, 307, 326
GeoGebra	277, 286, 293
geometry	
differential geometry	267
Euclidean geometry	90, 222-224, 226, 247, 250, 253, 255, 263, 268, 269, 274, 293
hyperbolic geometry	230
non-Euclidean geometry	222-224, 226, 247, 250, 252, 253
spherical geometry	229, 230
illustration aids	70, 72, 73, 85, 254, 260, 261
image denoising	235, 239, 241, 243
image of mathematics	20, 41, 55, 192
image processing	221-223, 227, 232-235, 239-241, 244-247, 333
imagination	6, 35, 53, 70
incubation	13, 19, 20, 50
independent learner	55, 192
inductive process	67-69, 85
inner motivation	192
inquiry	7, 33, 37, 45, 52, 53, 71, 74, 75, 85, 89, 90, 93, 191, 193, 198, 201, 216, 252, 254, 310
instruction	33-35, 39, 40, 47-49, 52, 53, 74, 162, 191

Subject index	**Page**

CPSIA information can be obtained
at www.ICGtesting.com
Printed in the USA
BVHW050951230322
632180BV00002B/68

9 789811 218743